T0318814

Thermodynamics

There are no such things as applied sciences,
only applications of science
Louis Pasteur (11 September 1871)

Dedicated to my wife, Anne, without whose unwavering
support, none of this would have been possible.

Industrial Equipment for Chemical Engineering Set

coordinated by
Jean-Paul Duroudier

Thermodynamics

Jean-Paul Duroudier

ELSEVIER

First published 2016 in Great Britain and the United States by ISTE Press Ltd and Elsevier Ltd

ISTE Press Ltd
27-37 St George's Road
London SW19 4EU
UK

Elsevier Ltd
The Boulevard, Langford Lane
Kidlington, Oxford, OX5 1GB
UK

www.iste.co.uk

www.elsevier.com

Notices

Knowledge and best practice in this field are constantly changing. As new research and experience broaden our understanding, changes in research methods, professional practices, or medical treatment may become necessary.

Practitioners and researchers must always rely on their own experience and knowledge in evaluating and using any information, methods, compounds, or experiments described herein. In using such information or methods they should be mindful of their own safety and the safety of others, including parties for whom they have a professional responsibility.

To the fullest extent of the law, neither the Publisher nor the authors, contributors, or editors, assume any liability for any injury and/or damage to persons or property as a matter of products liability, negligence or otherwise, or from any use or operation of any methods, products, instructions, or ideas contained in the material herein.

For information on all our publications visit our website at http://store.elsevier.com/

British Library Cataloguing-in-Publication Data
A CIP record for this book is available from the British Library
Library of Congress Cataloging in Publication Data
A catalog record for this book is available from the Library of Congress
ISBN 978-1-78548-176-5

Printed and bound in the UK and US

Contents

Preface

The observation is often made that, in creating a chemical installation, the time spent on the recipient where the reaction takes place (the reactor) accounts for no more than 5% of the total time spent on the project. This series of books deals with the remaining 95% (with the exception of oil-fired furnaces).

It is conceivable that humans will never understand all the truths of the world. What is certain, though, is that we can and indeed must understand what we and other humans have done and created, and, in particular, the tools we have designed.

Even two thousand years ago, the saying existed: "faber fit fabricando", which, loosely translated, means: *"c'est en forgeant que l'on devient forgeron"* (a popular French adage: *one becomes a smith by smithing*), or, still more freely translated into English, "practice makes perfect". The "artisan" (faber) of the 21st Century is really the engineer who devises or describes models of thought. It is precisely that which this series of books investigates, the author having long combined industrial practice and reflection about world research.

Scientific and technical research in the 20th century was characterized by a veritable explosion of results. Undeniably, some of the techniques discussed herein date back a very long way (for instance, the mixture of water and ethanol has been being distilled for over a millennium). Today, though, computers are needed to simulate the operation of the atmospheric distillation column of an oil refinery. The laws used may be simple statistical correlations but, sometimes, simple reasoning is enough to account for a phenomenon.

Since our very beginnings on this planet, humans have had to deal with the four primordial "elements" as they were known in the ancient world: earth, water, air and fire (and a fifth: aether). Today, we speak of gases, liquids, minerals and vegetables, and finally energy.

The unit operation expressing the behavior of matter are described in thirteen volumes.

It would be pointless, as popular wisdom has it, to try to "reinvent the wheel" – i.e. go through prior results. Indeed, we well know that all human reflection is based on memory, and it has been said for centuries that every generation is standing on the shoulders of the previous one.

Therefore, exploiting numerous references taken from all over the world, this series of books describes the operation, the advantages, the drawbacks and, especially, the choices needing to be made for the various pieces of equipment used in tens of elementary operations in industry. It presents simple calculations but also sophisticated logics which will help businesses avoid lengthy and costly testing and trial-and-error.

Herein, readers will find the methods needed for the understanding the machinery, even if, sometimes, we must not shy away from complicated calculations. Fortunately, engineers are trained in computer science, and highly-accurate machines are available on the market, which enables the operator or designer to, themselves, build the programs they need. Indeed, we have to be careful in using commercial programs with obscure internal logic which are not necessarily well suited to the problem at hand.

The copies of all the publications used in this book were provided by the *Institut National d'Information Scientifique et Technique* at Vandœuvre-lès-Nancy.

The books published in France can be consulted at the *Bibliothèque Nationale de France*; those from elsewhere are available at the British Library in London.

In the in-chapter bibliographics, the name of the author is specified so as to give each researcher his/her due. By consulting these works, readers may gain more in-depth knowledge about each subject if he/she so desires. In a reflection of today's multilingual world, the references to which this series points are in German, French and English.

The problems of optimization of costs have not been touched upon. However, when armed with a good knowledge of the devices' operating parameters, there is no problem with using the method of steepest descent so as to minimize the sum of the investment and operating expenditure.

A Logical Presentation
of Thermodynamics

1.1. Concepts in thermodynamics

1.1.1. *States of aggregation of material*

There are four states of aggregation: gas, liquid, solid and plasma.

In the discussions herein, we shall only consider the first three.

1.1.2. *Phases and components*

A material is said to be homogeneous if all of its properties vary contiguously. It is said to be uniform if all its properties have the same value throughout.

A homogeneous piece of material constitutes what is called a "phase". The number of phases that may be found in nature is, if not limitless, at least very large, because there are a considerable number of chemical species of which they may be composed.

A piece of material is said to be heterogeneous if its properties exhibit discontinuities. However, within a heterogeneous ensemble, there are homogeneous domains which make up phases.

An ensemble – be it homogeneous or heterogeneous – is often said to be either single- or poly-phase.

The terms "component" and "element" denote any species involved in the composition of a homogeneous or heterogeneous mixture, whose quantity may vary independently of the other components. The proportions of the different components of a mixture are often defined in concentrations (kmol.m^{-3} or kg.m^{-3}) or in fractions, also known as titers (molar, gravimetric or volumetric).

1.1.3. Thermodynamic system: variables and functions of state

A thermodynamic system is a material ensemble whose state is defined by so-called variables of state (composition, pressure, temperature, etc.). These variables are independents. The other variables defining the state are the "functions of state", which are obtained from the variables of state by using the "equations of state".

Instead of "system", we sometimes employ the term "physical agent".

1.1.4. Variables and intensive or extensive functions (parameters)

We distinguish between:

– intensive parameters, which do not depend on the size of the physical agent (pressure, temperature, titers, etc....),

– extensive parameters, whose value is proportional to the size of the system (mass, number of particles or molecules, energy, volume, etc.).

1.1.5. Isolated, open and closed systems

An isolated system is a system which exchanges nothing with its environment, and whose boundary is immovable.

A system is open or closed depending on whether or not it exchanges material with the outside world.

In statistical physics, isolated systems are described by the micro-canonical ensemble. A closed system (which exchanges "energy but not material with the outside world") is described by the canonical ensemble. A system which exchanges both energy and material with its environment is described by the large canonical ensemble.

1.1.6. *Energy and evolution of a system. The concept of work*

Essentially, in thermodynamics, we study the consequences in terms of energy of the changes occurring in a system, drawing inspiration from elementary mechanics (the mechanics of simple machines). We characterize the work created from a potential energy by the product of a generalized force X_i by a generalized displacement x_i. The conjugate values X_i and x_i are linked by an equation of state, and there are as many equations of state as there are types of work. Conventionally, the generalized forces are counted positively, and so too is any work exerted by the outside on the system – i.e. the work received by the system. This gives us the sign of the generalized displacement.

1.1.7. *Classification of systems*

For our purposes here, the following classification is used:

– depending on the number of different types of work exchanged, the system may be:

- simple if it involves only one type of work,

- complex if it involves several different types of work, which may, potentially, influence one another.

– depending on the number of components, the system is:

- unitary if it contains only one component,

- multiple if it is a mixture of several components.

1.1.8. *The concept of a "reservoir"*

A reservoir, as understood in thermodynamics, is a system whose mass is infinite (however large we want it to be). A finite injection of material does not alter the mass of the reservoir or the properties for which it was chosen.

For instance, we may agree that the pressure and/or temperature of the reservoir are invariable when it exchanges energy with a finite system. In this case, the reservoir shall be said to be isobaric and/or isothermal. The idea of a reservoir can be extended to apply to any generalized force.

1.2. Tenets of thermodynamics

1.2.1. *First tenet: concept of equilibrium*

"An isolated system is either in a state of equilibrium which it cannot spontaneously abandon, or else it tends toward that state".

Out of all the possible states in which a system may find itself, the state of equilibrium is the most probable.

When two systems are in contact with one another and at equilibrium, the generalized forces and the temperature of the two systems are equal. The result of this is that their effects on the boundary between the two systems are neutralized.

Hence, within a homogeneous system at equilibrium, the generalized forces have the same value everywhere. The system is uniform.

1.2.2. *Second tenet: transitivity of equilibrium*

The second postulate can be stated thus:

"If the two systems A and B are separately at equilibrium with a third system C, then systems A and B are at equilibrium with one another".

In particular, the property of transitivity enables us to compare the temperature of two systems without their being in direct contact, with the system C acting as a thermometer. We say that the two systems are at thermal equilibrium if their temperatures are equal. The same is true for the pressures, measured with a manometer.

1.3. First law of thermodynamics and internal energy

1.3.1. *Concepts of hot and cold: empirical temperature*

Since the dawn of time, human beings have always been able to distinguish hot (e.g. fire) from cold (e.g. snow or ice). The sensitivity of touch means we can discern whether, say, one body of water is hotter than another, and is therefore colder.

This distinction gives us the concept of empirical temperature, which we can associate, for instance, with the thermal expansion of a gaseous, liquid or even solid substance, and attach a numerical value to the measurement. *It has been agreed through convention that, the hotter a substance is, the higher its temperature is.*

It was then remarked upon that, if we bring a hot substance and a cold substance into contact, the hot substance cools down and the cold substance warms up, and ultimately, the two substances come together at an intermediate temperature. This gave us the concept of heat, whereby the hotter a substance is, the more heat it carries. The important property of heat is that it moves from the hot substance into the cold substance so that the initial difference in empirical temperature tends to even out. This is confirmed by experience.

1.3.2. *Consequence: evaluation of heat: specific heat capacity*

It was decided that the "heat" of a mass M should be quantified by multiplying M by its empirical temperature t:

$$Q=Mt$$

In the absence of mechanical operations, the amount of heat present is conserved. For instance, if a cold solid substance of mass M is immersed in a quantity of hot water of the same mass M, *we see that* the solid warms up, and its temperature increases by Δt_S; meanwhile, that of the water decreases by Δt_E. The conservation of heat is formulated as:

$$\delta Q=M\Delta t_E=MC_S\Delta t_S$$

From this equation, we can deduce the value of the specific heat capacity C_S of the solid (with that of water being taken as 1). Every substance has a specific heat capacity which comes into play in all "noticeable" heat exchanges – i.e. heat exchanges without a change of state (the vaporization of a liquid is an example of a change of state). Thus, it becomes possible to evaluate the amount of heat passed to any given substance by measuring the variation in its empirical temperature.

1.3.3. *Equivalences of energies*

Joule's experiments showed that, if work is expended to produce heat by friction, that heat δQ is proportional to the work δW carried out, and the coefficient of proportionality has a universal value.

If we define the scale of empirical temperature as being proportional to the expansion of a gas at evanescent pressure with $t = 0$ for ice at equilibrium with water and 100 for water at equilibrium with its vapor (steam) in an absolute atmosphere, the unit of heat shall be the kilocalorie, which corresponds to a rise in temperature from 14 to 15 degrees in 1 kilogram of water. The unit of work is the joule, which is measured in $kg.m^2.s^{-2}$. The conversion coefficient is such that:

   1 kilocalorie=4187 joules

1.3.4. *External energy of a system*

The external energy is the sum of:

– the potential energy of the system due to an external force field; and

– the kinetic energy of the system when it moves as a whole.

1.3.5. *General expression of work*

By definition, the work is the product of a force by the displacement of its point of application, provided that displacement is equipollent to the force.

In thermodynamics, it is possible to generalize this definition, as shown in Table 1.1, where the product of the two columns gives the energy received by the system.

Type of work	Generalized force	Generalized displacement
Resistance of a spring to compression	F	-dL
Resistance of a thread to traction	F	dL
Tension of a film on the surface	σ	$d\Sigma$
Pressure of compression	P	-dV
Chemical potential of a kilomole	μ	dn

Table 1.1 *Generalized forces and displacements*

We can also envisage the work of an electrical field on a charge or that of a magnetic field on a current, but will not do so here.

Let the displacements x_i take place in the same direction as the force X_i, with this force being directed from the inside of the system toward the outside. The elementary work exerted by the system on its external environment is:

$$\delta W_i = X_i dx_i$$

This corresponds to energy received by the system:

$$W_i = -\int_{x_0}^{x} X_i dx_i + \text{const.}$$

We use the term "pneumatic work" to speak of the work involving the pressure and the volume, even if that work involves a liquid rather than a gas. The corresponding potential energy received is:

$$W_P = -\int_{V_0}^{V} PdV$$

Note that these energies are *very specific to the problem at hand*, because the value of the integral depends on an additional variable – the temperature – which we can vary at will and, generally speaking, the W_i values depend on the particular stresses imposed when calculating the integral. This shows that the W_i values *are not variables of state*. This shall be explained later on. For now, we can simply write the differential expression of the internal energy U:

$$dU = \delta Q - \sum_i X_i dx_i$$

The variation of a system's internal energy is the sum of the heat and the work received by the system.

1.3.6. Introduction to enthalpy

The enthalpy is defined by:

$$H = U + PV$$

The advantage to adding PV to the internal energy is as follows.

Consider a fluid circulating in a piping system and, more specifically, a "slice" of fluid of a given mass. During the time period dt, the posterior face of the slice travels the distance dL and, if A is the area of the cross-section of the pipe, upstream the slice receives the work PAdL. Over the distance dL, the pressure of the fluid has varied by dP (dP < 0) because of friction. On its anterior face, the fluid has exerted the work $(P + dP)$ A $(dL + \delta L)$, where $A\delta L$ represents the variation in the volume of the fluid. The net variation in the internal energy of the slice of fluid – i.e. the energy received by the circulating fluid – is:

$$dU = APdL - (P + PD)A(dL + \delta L) = -(AP\delta L + AdPdL + AdP\delta L)$$

Let us posit:

$$AdL = V \text{ and } A\delta L = dV$$

Thus, if we ignore the second-order term:

$$dU = -(PdV + VdP) = -d(PV)$$

This term –d (PV) expresses the result of the flow of the fluid. It could be called the "circulation term". Note that:

$$dU + d(PV) = dH = 0$$

We can now see why enthalpy is so widely used in establishing heat balances for industrial units.

Generally, in this particular case, we can write:

$$dH = dU + d(PV) = \delta Q + VdP$$

Hence, we can say that the enthalpy quantifies the energy of a *circulating* fluid, which is highly useful for the industry of processes working continuously and in a permanent regime.

More generally, though, whether an isenthalpic system is circulating or static, any variation in the product PV is compensated by an opposing variation in the internal energy.

NOTE.– The enthalpy and internal energy of a perfect gas depend only on the temperature. However, in the case of real fluids, these two values also depend on the pressure.

1.3.7. *Heat is not a function of state*

If the system (e.g. a fluid) is made to undergo a cycle, once it has returned to its initial state, the internal energy has not changed, and we can write:

$$\oint dU = \oint \delta Q + \oint \delta W = 0$$

The sum of the heat Q and the work W exchanged with the outside environment has not varied. However, during the course of the cycle, it is possible to:

– transform work into heat. This is the phenomenon of friction, which to a greater or lesser degree occurs in machines;

– trigger a chemical reaction or simply make a mixture (with the system's internal components) simply by opening a valve;

– subject the system to an arbitrary amount of work, which can be shown by the area enclosed by the curve representing the system's evolution, with the force and the displacement being used as coordinates (on condition, of course, that heat is applied or removed at the appropriate points in the cycle).

Put differently, the work carried out by system depends on the nature of the cycle, and consequently we have:

$$\oint dW \neq 0 \quad \text{and therefore} \quad \oint \delta Q = -\oint \delta W \neq 0$$

Given the arbitrary nature of the choice of cycle, there can be no function of state W, whose derivative defines the elementary work exchanged. In other words, the value W is not a function of state with an exact total derivative, so we write δW instead of dW and δQ instead of dQ.

Hereinafter, we shall adopt the following conventions:

– $\delta Q > 0$ for the heat received by the system;

– $\delta W > 0$ for the work exerted by the external environment *upon* the system – i.e. the work received by the system.

1.4. Second law and entropy

1.4.1. *Second law of thermodynamics*

We shall state this law as follows:

"If a thermally-homogeneous fluid receives heat, it cannot *entirely* transform that heat into work without any other exchange with its environment".

1.4.2. *Conversion of heat into work. Prohibited states*

Consider the following two transformations relating to a thermally-homogeneous fluid.

1) if a fluid receives work $\delta W_1 > 0$, it can fully transform that work into heat, which it expels toward the outside:

$$dU_F = \delta Q_1 + \delta W_1 = 0 \text{ where } \delta W_1 > 0 \text{ and } \delta Q_1 < 0$$

2) if a fluid receives heat $\delta Q_2 > 0$, it cannot fully transform that heat into work, and the balance heat $\delta Q_R < 0$ is restored to the outside.

$$dU_F = \delta Q_2 + \delta Q_R + \delta W_2 = 0 \text{ with } \delta Q_2 > 0 \; \delta Q_R < 0 \text{ and } \delta W_2 > 0$$

In other words, the work performed is:

$$-\delta W_2 = \delta Q_2 + \delta Q_R < \delta Q_2$$

Now consider the ensemble formed by the fluid and its environment. The first transformation of that system is irreversible because, if we set $\delta Q_2 = -\delta Q_1$, it is not possible to return to the initial state by transformation 2.

$$-\delta W_2 = \delta W_1 + \delta Q_R < \delta W_1$$

In other words, certain states are *forbidden* from a given state. In the present case, the initial state is forbidden from the final state. Transformation 1, therefore, is *irreversible*.

1.4.3. *Pfaff formula and holonomicity*

We know that the variation of the work energy can be put in the form:

$$\delta W = -\sum_i X_i dx_i \text{ (for instance: } X\_i = P \quad \text{and} \quad x\_i = V)$$

However, for an isolated system such as the ensemble of the fluid and its environment:

$$dU = \delta Q + \delta W = 0$$

Therefore:

$$\delta Q = dU + \sum_i X_i dx_i$$

δQ is a Pfaff formula. Indeed, the X_i values are not the partial derivative of Q, so δQ is not a total derivative. However, if we can find an "integrating factor" in the Pfaff formula, it will be said to be "holonomic", and we shall have:

$$\frac{\delta Q}{\lambda} = d\sigma \text{ or indeed } \delta Q = \lambda d\sigma$$

The factor $1/\lambda$ is the integrating factor of the variation δQ. We shall now see why this is possible.

1.4.4. *Carathéodory's theorem and existence of entropy*

Carathéodory's theorem is stated as follows:

"If, in the vicinity of the state P_0, there are states P that cannot be reached from P_0, along the curves that are solutions to the equation:

$$\sum_i X_i dx_i = 0$$

This property constitutes a necessary and sufficient condition for Pfaff's formula to be holonomic".

The demonstration of this theorem can be found in Honig [HON 95]. This demonstration is somewhat lengthy.

However, with regard to heat/work exchanges, we have seen the existence of forbidden states. Hence, the form δQ is holonomic, and we have:

$$\delta Q = \lambda d\sigma$$

The value σ is called the "empirical entropy".

1.4.5. Reversibility and entropy

We shall now verify that if we look for a function σ which remains invariant in any reversible (and adiabatic) transformation of a system, we find:

$$\delta Q = \lambda d\sigma$$

$-$ adiabatic means:

$$\delta Q = dU + \sum_i X_i dx_i = \frac{\partial U}{\partial t} dt + \sum_{i=1} \left[\frac{\partial U}{\partial x_i} + X_i \right] dx_i = 0$$

or indeed:

$$\delta Q = \sum_{i=1}^{n+1} Z_i dx_i = 0$$

$-$ reversible and adiabatic means:

$$d\sigma = \frac{\partial \sigma}{\partial t} dt + \sum_{i=1}^{n} \frac{\partial \sigma}{\partial x_i} dx_i = \sum_{i=1}^{n+1} Y_i dx_i = 0$$

As the Z_i are not the partial derivatives of a function of state Q (which does not exist), we have seen that the value δQ is a "Pfaff formula".

The two linear forms in dx_i must always reach the value of 0 simultaneously $-$ i.e. for the same values of x_i. Let us now show that these two forms must necessarily be proportional.

To begin with, let us vary two values x_1 and x_2: let dx_1 and dx_2 represent their variations:

$$\delta Q = Z_1 dx_1 + Z_2 dx_2 = 0 \text{ or indeed } Z_1 dx_1 = -Z_2 dx_2$$

Similarly, for σ:

$$d\sigma = Y_1 dx_1 + Y_2 dx_2 = 0 \text{ or indeed } Y_1 dx_1 = -Y_2 dx_2$$

If we divide member by member, we find:

$$\frac{Z_1}{Y_1} = \frac{Z_2}{Y_2} = \lambda = \frac{Z_1 dx_1}{Y_1 dx_1} = \frac{Z_2 dx_2}{Y_2 dx_2} = \frac{Z_1 dx_1 + Z_2 dx_2}{Y_1 dx_1 + Y_2 dx_2}$$

Let us bring one more variable into play, with an arbitrary value of dx_3, we should have:

$$Z_1 dx_1 + Z_2 dx_2 = -Z_3 dx_3$$

$$Y_1 dx_1 + Y_2 dx_2 = -Y_3 dx_3$$

Divide member by member; we have:

$$\lambda = \frac{Z_3}{Y_3} = \frac{Z_1 dx_1 + Z_2 dx_2 + Z_3 dx_3}{Y_1 dx_1 + Y_2 dx_2 + Y_3 dx_3}$$

This can immediately be generalized to $n + 1$ variable.

Thus, we confirm the result obtained by Carathéodory's theorem.

$$\delta Q = \lambda d\sigma$$

Remember, σ is the "empirical entropy".

Conversely, if a non-reversible transformation affects an isolated system, we shall show in section 1.6.5 that a transformation is always accompanied by an increase in entropy.

1.4.6. *Entropy: an extensive variable*

Consider two systems A and B characterized by the variables of state x_i for the system A and x_j for the system B. Also consider the system E which is an ensemble of A and B. Suppose that A, B and therefore E exchange heat reversibly with an isothermal reservoir R. Reversibility implies that the empirical temperature t must be the same everywhere – i.e. for A, B, E and R.

The heat δQ_E received by E is the sum of the heats received by A and B:

$$\delta Q_E = \delta Q_A + \delta Q_B$$

Thus, for the total derivative of σ_E:

$$d\sigma_E = \frac{\lambda_A}{\lambda_E} d\sigma_A + \frac{\lambda_B}{\lambda_E} d\sigma_B \qquad [1.1]$$

Because t, x_i and x_j are not explicitly involved in equation [1.1], σ_E is independent of those variables, and the same is true of the ratios λ_A/λ_E and λ_B/λ_E, which are respectively equal to $\partial\sigma_E/\partial\sigma_A$ and $\partial\sigma_E/\partial\sigma_B$. In addition, λ_E cannot depend on the values of x_i, whereas λ_B must be dependent on them in order for λ_B/λ_E not to be. However, λ_B cannot depend on the variables of the system B. Similarly, λ_E does not depend on x_j and on λ_A.

The x_i and x_j values do not appear in λ_E. The result of this is that λ_A cannot depend directly on x_i nor λ_B on x_j.

The above does not involve the empirical temperature t, which is the same between all three 3 systems A, B and E, so that the three functions λ are of the form:

$$\lambda_A(\sigma_A, t), \lambda_B(\sigma_B, t) \text{ and } \lambda_E(\sigma_A, \sigma_B, t)$$

Yet the temperature t does not appear to be an independent and explicit variable in $d\sigma_E$. Therefore, it is necessary for the λ to be of the following form:

$$\lambda_A = \psi(t)f_A(\sigma_A) \; \lambda_B = \psi(t)f_B(\sigma_B) \; \lambda_E = \psi(t)f_E(\sigma_A, \sigma_B)$$

and, consequently:

$$\delta Q_A = \psi(t)f_A(\sigma_A)d\sigma_A \; \delta Q_B = \psi(t)f_B(\sigma_B)d\sigma_B$$

$$\delta Q_E = \psi(t)f_E(\sigma_A, \sigma_B)d\sigma_E = \delta Q_A + \delta Q_B = \psi(t)[f_A(\sigma_A)d\sigma_A + f_B(\sigma_B)d\sigma_B]$$

We set:

$$f_A(\sigma_A)d\sigma_A = dS_A \qquad f_B(\sigma_B)d\sigma_B = dS_B \text{ and } f_E(\sigma_A, \sigma_B)d\sigma_E = dS_E$$

This gives us the additive relation:

$$dS_E = dS_A + dS_B$$

Let us integrate with the condition that those three values reach 0 at the same time. We obtain:

$$S_E = S_A + S_B$$

We use the term *metric entropies* or, more simply, "entropies", to speak of these three values. This last relation demonstrates that the entropy is an extensive property.

Now let us set:

$$\psi(t) = T$$

T is the thermodynamic temperature. We can now write:

$$dU = \delta Q + \delta W = TdS - PdV$$

$$dH = TdS + VdP$$

1.4.7. *Sign of the absolute temperature [BAZ 83]*

The derivative of the entropy can be written:

$$dS = \frac{\left(\frac{\partial U}{\partial T}\right)_{x_i} dT + \sum_i \left[\left(\frac{\partial U}{\partial x_i}\right)_T + X_i\right]dx_i}{T} = \left(\frac{\partial S}{\partial T}\right)_{x_i} dT + \sum_i \left(\frac{\partial S}{\partial x_i}\right)_T dx_i$$

Thus:

$$\left(\frac{\partial S}{\partial T}\right)_{x_i} = \frac{1}{T}\left(\frac{\partial U}{\partial T}\right)_{x_i} \quad \text{and} \quad \left(\frac{\partial S}{\partial x_i}\right)_T = \frac{1}{T}\left[\left(\frac{\partial U}{\partial x_i}\right)_T + X_i\right]$$

The cross derivatives of S are such that:

$$\frac{\partial}{\partial x_i}\left[\frac{1}{T}\left(\frac{\partial U}{\partial T}\right)_{x_i}\right] = \frac{\partial}{\partial T}\left\{\frac{1}{T}\left(\frac{\partial U}{\partial x_i}\right)_T + X_i\right\}$$

Hence:

$$T = \left(\frac{\partial X_i}{\partial T}\right)_{x_i} = \left(\frac{\partial U}{\partial x_i}\right)_T + X_i$$

However, experience tells us that there is a biunivocal correspondence between the empirical temperature t and the thermodynamic temperature T.

Consequently:

$$T = \psi(t) \text{ is equivalent to } t = \varphi(T)$$

Hence:

$$T\left(\frac{\partial X_i}{\partial T}\right)_{x_i} = T\left(\frac{\partial X_i}{\partial t}\right)_{x_i}\frac{dt}{dT} = \left(\frac{\partial U}{\partial x_i}\right)_T + X_i$$

Note that the measurement of an intensive value X_i (here with the generalized force being different from the temperature) as a function of the empirical temperature t merely acts as a thermometer. Finally:

$$\int_{T_0}^{T}\frac{dT}{T} = \int_{t_0}^{t}\frac{(\partial X_i/\partial T)_{x_i}}{(\partial U/\partial x_i)_T + X_i}\,dt = I$$

Hence:

$$T = T_0 e^1$$

Thus, the thermodynamic temperature has an invariable sign. For simplicity's sake, the choice which was made is:

$$T_0 > 0 \quad \text{and therefore} \quad T > 0$$

1.4.8. Thermodynamic temperature scale

The experience shows that, for a pressure that is evanescent but different to zero, all gases have the same volume at the same temperature. In addition, statistical physics theoretically establishes that the perfect gases obey the universal relation:

$$V = \frac{R}{P}T$$

The perfect gas is the limit case of real gases for an evanescent pressure.

In addition, it has been accepted that there is a gap of 100 K (K is the Kelvin; the unit of absolute temperature) between the freezing point of water T_0 and its boiling point T_{100}, both taken at one absolute atmosphere. We can write:

$$\frac{V_0}{T_0} = \frac{V_{100}}{T_0+100} = const.$$

and experience shows us that:

$$T_0 = 273.15 \text{ K}$$

Hence, the temperature T is a universal value.

1.5. Gibbs energy and Euler's theorem

1.5.1. Chemical potential [GIB 99]

Consider a fluid F in which a molecule A and a molecule B combine to form the compound C by a spontaneous athermic reaction. This reaction is supposed to take place at constant T and P.

The approach and combination of the two reagents corresponds to a *minimum* of what could be called the "potential energy of the reagents", to borrow an image from mechanics:

$$g_C < g_A + g_B$$

This potential energy is not an intrinsic property of reagents A and B because, first and foremost, it is dependent on the composition of the fluid, and therefore on its nature, on the pressure and the temperature. The energy of each reagent is what could be called a marginal energy which corresponds to the variation of the overall energy of the fluid for the appearance or disappearance of a kilomole of reagent whose index is i. If g_F is the energy of the fluid:

$$dG_F = \left(\frac{\partial G_F}{\partial n_A}\right)_{T.P} dn_A = g_A dn_A = \mu_A dn_A$$

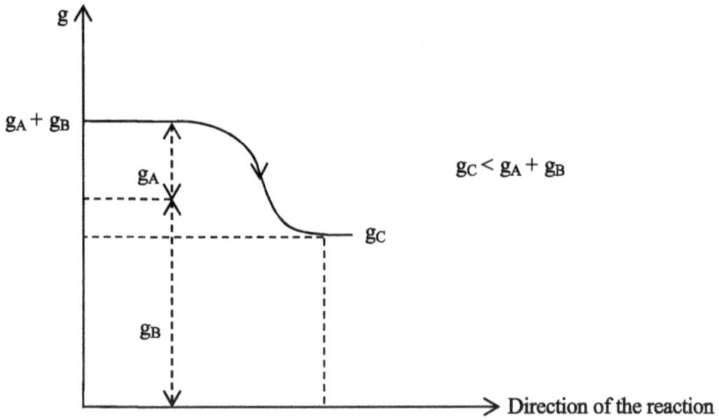

Figure 1.1. *Energy balance of a reaction*

The term dn_A is positive if we add kilomoles of A to the fluid and negative for the inverse operation. In relation to the fluid, the potential energy balance is written simply:

$$\Delta a_F = \mu_C dn_C - \mu_A dn_A - \mu_B dn_B < 0$$

The μ are simply called the "chemical potentials" and are measured in joules per kilomole.

More generally, any chemical reaction can be written:

$$\sum_i v_i M_i = 0$$

The v_i are the stoichiometric coefficients of the reaction, generally counted positively for the reagents and negatively for the products. The M_i are the symbols of the molecules involved in the reaction. Finally, the reaction is expressed by:

$$\Delta G = G_{products} - G_{reagents} = -\sum v_i \mu_i < 0$$

The energy G characterizes the fluid, as we have seen. It is an energy available (free) for the reactions (which, for the time being, we suppose to be athermic) occurring within that fluid. However, we know that the energy of a circulating fluid is characterized by its enthalpy, so that, quite naturally, the energy G has been called the "Gibbs energy", referring to the fraction of the enthalpy which is available for chemical reactions (and also for the transfers from one phase to the other, as we shall see). The symbol G was chosen in honor of the thermodynamics expert Gibbs, who first introduced this concept.

Thus, consider a fluid which is constituted gradually by accumulating kilomoles of its components, with the composition always being kept constant. Hence, if n_T is the total number of kilomoles of the fluid at a given time, and if the x_i characterize the molar fractions, then at each time τ, we have:

$$\frac{dn_i}{d\tau} = x_i \frac{dn_T}{d\tau}$$

With the temperature, pressure and composition remaining constant, the chemical potentials remain constant as well, and the Gibbs energy G is written:

$$G = \sum_i \mu_i n_i = \sum_i \mu_i x_i n_T$$

(at given levels of T and P).

1.5.2. *Properties of the chemical potential*

Finally, the chemical potential is the energy which eliminates a kilomole when it leaves a fluid. Conversely, and less easy to imagine, a kilomole which joins a fluid corresponds to a variation of the Gibbs energy of that fluid which is equal, not to the chemical potential which it had in the original fluid, but to the chemical potential corresponding to the destination fluid (with its temperature, its pressure and its composition). We again see that the chemical potential is not an intrinsic property of each component.

Ultimately, when fluid 1 loses a kilomole of component i, it loses the Gibbs energy μ_{1i}, and when fluid 2 receives the same, it gains the Gibbs energy μ_{2i} and, for dn_i kilomoles, the variation in the Gibbs energy of the system that is the union of the two fluids is:

$$dG = (\mu_{2i} - \mu_{1i})dn_i$$

What has become of that energy? We shall see the answer to this later on. Right now, though, we know that as the Gibbs energy is a collective *potential* for all the components of a system (rather than a single fluid), the spontaneous *transfer* from phase 1 to 2 results in a *decrease in the Gibbs energy* of the overall system.

Consider a system which can exchange energy with the environment. We know that its internal energy varies in the following manner:

$$dU = TdS - PdV$$

If, in addition, it is an open system – i.e. one that can also exchange material with the outside world – and if we suppose that the system exchanges dn_i kilomoles of the component i, then its total energy (its internal energy) will change from $\mu_i\, dn_i$. *As before*, the μ_i will be determined by the temperature, pressure and composition of the system, and we can write:

$$dU = TdS - PdV + \sum_i \mu_i dn_i$$

Here, though, we have:

$$\mu_i = \left(\frac{\partial U}{\partial n_i}\right)_{S,V} \quad \text{instead of} \quad \mu_i = \left(\frac{\partial G}{\partial n_i}\right)_{T,P}$$

Note that the same chemical potential expresses the variation in internal energy at constant S and V and the variation in Gibbs energy at constant T and P.

1.5.3. *Euler's theorem*

This equation links the entropy and the Gibbs energy by way of the enthalpy. Indeed, we shall show that:

$$H = TS + \sum_i \mu_i n_i = TS + G$$

Consider a system whose internal energy is:

$$U^0\left(S^0, V^0, n_i^0, \dots n_k^0\right)$$

We know that:

$$dU = TdS - PdV + \sum_{i=1}^{k} \mu_i dn_i \tag{1.2}$$

Let us give the same small relative increase to the k numbers of moles n_i^0, with the rate of increase being *the same* for all these variables.

$$n_i^1 = (1+\varepsilon)n_i^0 \quad \forall i$$

However, the internal energy is an extensive variable that is proportional to the numbers of moles n_i. The same is true for S and V, so:

$$U^1\left(S^1, V^1, n_i^1\right) = (1+\varepsilon)U^0\left(S^0, V^0, n_i^0\right), \text{ meaning that } {}'U^1 = (1+\varepsilon)U^0$$

where:

$$S^1 = (1+\varepsilon)S^0 \text{ and } V^1 = (1+\varepsilon)V^0$$

Let us expand U into a Taylor series:

$$U^1 = U^0 + \frac{\partial U}{\partial S}\varepsilon S^0 + \frac{\partial U}{\partial V}\varepsilon V^0 + \cdots + \frac{\partial U}{\partial n_k}\varepsilon n_k^0$$

In light of equation [1.2]:

$$\frac{\partial U}{\partial S} = T \qquad \frac{\partial U}{\partial V} = -P \qquad \left(\frac{\partial U}{\partial n_i}\right)_{S,V} = \mu_i$$

Hence:

$$U^1 = U^0 + \varepsilon\left(TS^0 - PV^0 + \sum_i \mu_i n_i^0\right)$$

However:

$$U^1 - U^0 = \varepsilon U^0$$

Therefore:

$$U^0 = TS^0 - PV^0 + \sum_i \mu_i n_i^0$$

This means that:

$$H^0 = TS^0 + G^0 \quad \text{(Euler's equation)}$$

At absolute zero, statistical physics shows that the entropy is null. The internal energy is too.

In reality, for instance for a harmonic linear oscillator, quantum mechanics shows that, at absolute zero, there remains a residual movement and that, consequently, the internal energy is not reduced to zero (Guinier [GUI 49], p. 59). However, for our purposes here, we shall make the hypothesis that $U = 0$ for $T = 0$ K. With regard to the enthalpy, such an assumption is not necessary, because that value is generally not used, except in the heat balances where only the differences play a part. In addition, the volume does not disappear at absolute zero.

1.6. Values of state. Thermodynamic potentials

1.6.1. Definitive form of the values of state

We can now construct Table 1.2.

Name	Definition	Derivative
Internal energy	$U\,(S, V, n_i)$	$dU = TdS - PdV + \sum_i \mu_i dn_i$
Enthalpy	$H = U + PV$	$dH = TdS + VdP + \sum_i \mu_i dn_i$
Gibbs energy	$G = H - TS$	$dG = -SdT + VdP + \sum_i \mu_i dn_i$
Helmholtz energy	$F = U - TS$	$dF = -SdT - PdV + \sum_i \mu_i dn_i$
Grand potential	$A = -PV = U - ST - \sum n_i \mu_i$	$dA = -SdT - PdV - \sum n_i d\mu_I$

Table 1.2. Definitive form of the values of state

We use the grand potential to calculate the functions of state of perfect gases (see section 2.1.2).

We can use the term "generative functions" for the five functions of state in Table 1.2, because their partial derivatives enable us to obtain the variables T, P, S, μ_i and even n_i and V.

The total derivative of a generator function is the sum of products of the form $Y_i dZ_i$. The two values Y_i and Z_i are called *conjugate values*.

The *natural variables* of a generator value are those whose derivative appears in the total derivative of the generator. For example:

$$dU = TdS - PdV + \sum_i \mu_i dn_i$$

The values S, V and n_i are the natural variables of the internal energy U. The values T, P and the chemical potentials μ_i are the partial derivatives of the generator function U:

$$\left(\frac{\partial U}{\partial S}\right)_{V,n} = T \quad \left(\frac{\partial U}{\partial V}\right)_{S,n} = -P \quad \left(\frac{\partial U}{\partial n_i}\right)_{S,V} = \mu_i$$

1.6.2. *Irreversibility*

Let us orientate the generalized displacements in the direction of the generalized forces. When a displacement takes place between two systems A and B in contact with one another, the gain of one is equal to the loss of the other:

$$dx_A + dx_B = 0$$

The generalized displacement takes place from the system where the force is greater (X_A) toward the system where the force is lesser (X_B). The resulting force X_R can be deduced from this:

$$X_R = X_A - X_B > 0 \quad (X_A > 0 \; X_B > 0)$$

The work exchanged by system A is negative:

$$\delta W_A = -X_A dx_A < 0 \quad \text{(we adopt } dx_A > 0)$$

The work gained by system B is positive:

$$\delta W_B = -X_B dx_B > \quad 0 \; dx_B < 0$$

(The resistant force X_B is positive).

Now consider the system R which is the union of systems A and B. The resultant work is δW_R:

$$\delta W_R = \delta W_A + \delta W_B = -X_A dx_A - X_B dx_B$$

Let:

$$\delta W_R = -(X_A - X_B) dx_A < 0$$

This work was lost in the operation. It could be reduced if we decrease the difference between X_A and X_B and, for a difference of zero, there will no longer be any loss, but the operation would take an infinite amount of time. The operation would then be reversible, although it was irreversible for $X_A > X_B$ (in order to compensate the inevitable friction linked to the displacements).

More specifically, an exchange is said to be reversible if:

$$X_A > X_B \text{ However, } X_A - X_B < \varepsilon$$

Here, ε is a positive number less than any value given in advance, however small it needs to be. This condition means that the state of two systems does not need to be modified if we reverse the direction of the exchange, which justifies the use of the term "reversible".

Finally, we can say:

"The faster an exchange of work takes place, the more irreversible it is. On the other hand, a reversible exchange requires an infinite amount of time".

1.6.3. *Entropy increases when an internal stress is relieved*

Consider systems which are each made up of two reservoirs, separated by:

– a calorifugic membrane (impermeable to heat);

– a membrane impermeable to crossing by the material;

– the declutching of coupling between a motor and machine.

The relieving of such a stress immediately leads to a spontaneous, irreversible transformation which results in an *increase in entropy*. The derivative of the internal energy of the system R, which is the union of the two reservoirs (the system R is supposed to be *isolated*), is:

$$dU = TdS + \delta W + \sum_i \mu_i dn_i = 0 \text{ (where } \delta W \text{ can be } -PdV)$$

We shall envisage three situations, and in each, write that the system R loses what reservoir 1 loses and gains what reservoir 2 gains after the relieving of the stress. In each of the three systems, there is an immediate transfer between the two reservoirs as soon as the stress is removed.

1) Transfer of heat Q from reservoir 1 at the constant temperature T_1 to reservoir 2 at the constant temperature $T_2 < T_1$. The variation in entropy for the overall system R is:

$$dS = dS_1 + dS_2 = \frac{-S}{T_1} + \frac{Q}{T_2} = Q\left(\frac{1}{T_2} - \frac{1}{T_1}\right) > 0$$

2) Transfer of energy by an ordinary machine fed by reservoir 1 (electrically, for example), receiving the work W_1 from that reservoir and carrying out the work $W_2 < W_1$ in reservoir 2. The two reservoirs are supposed to be at the same temperature.

$$dU = TdS + W_2 - W_1 = 0$$

$$dS = \frac{W_1 - W_2}{T} > 0$$

3) Transfer of material dn_i between two reservoirs at the same temperature and pressure, but with different compositions and, therefore, such that $\mu_{1i} > \mu_{2i}$.

$$dU = TdS + (\mu_{2i} - \mu_{1i})dn_1 = 0$$

$$dS = \frac{(\mu_{1i} - \mu_{2i})dn_i}{T} > 0$$

1.6.4. *Generalization of the notion of entropy*

We have introduced the concept of entropy with regard to the heat transfer or certain irreversible transformations do not lead to the creation of heat. Consider, for example, an irreversible athermic and isochoric reaction taking place in an isolated system.

In light of section 1.6.1 and Table 1.2:

$$dF = -SdT - PdV + \sum_i \mu_i dn_i$$

As the reaction is athermic and isochoric:

$$dT = 0 \quad \text{and} \quad dV = 0 \text{ so therefore } dF = \sum_i \mu_i dn_i$$

Still according to section 1.6.1:

$$dU = TdS - PdV + \sum_i \mu_i dn_i = TdS + dF$$

As the system is isolated:

$$dU = 0 \quad \text{so} \quad dS = -\frac{dF}{T}$$

In the same way that, at constant T and P, a chemical reaction is spontaneous only if $\Delta G < 0$, similarly, at constant T and V, a spontaneous reaction is only possible if $\Delta F < 0$. The result of this is that, for the reaction in question, $\Delta S > 0$.

In the general case of irreversibility dissipating the energy δE, it is easy to see that δE is a Pfaff form (see section 1.4.3):

$$\delta E = \sum_i X_i dx_i$$

In addition, in an irreversible transformation, the initial state has become inaccessible. Carathéodory's theorem thus applies, so the form δE is holonomic and can be written:

$$\delta E = \varphi(t)d\Sigma$$

We shall call the value Σ *entropy*, and represent it hereinafter by the letter S. By an approach identical to that adopted for the *thermal* entropy (see sections 1.4.3 to 1.4.6), we set:

$$\varphi(t) = T$$

We simply need to make $\delta E = \delta Q$ to verify the consistency of the two approaches.

NOTE.– A fluid circulating in a pipe generates friction on the walls. The energy thus dissipated (in the form of heat) is carried by the fluid, which increases its entropy and, simultaneously, the pressure of the fluid decreases, and we have:

$$dH = TdS + VdP = 0 \text{ where } dS > 0 \text{ and } dP < 0$$

1.6.5. *Monotonic variation in entropy*

We shall begin by showing that any value S of the entropy is a boundary (an extremity, a limit) of the interval R which supports the values accessible from S by an irreversible adiabatic transformation. Indeed, if this proposition was not true, it would be impossible, in the vicinity of S, to find a forbidden state, and Carathéodory's theorem would no longer apply to the function S.

Now suppose that, in the wake of an irreversible adiabatic transformation, the entropy has increased from $S_{initial}$ to the value S_{final}. We shall show that in such an evolution, it is impossible to *find* a value $S_1 > S_{initial}$ such that an irreversible adiabatic transformation leads to a value S_2 lower than S_1. Indeed, if this were the case, there would be a common interval between, firstly, the interval $[S_{initial}, S_{final}]$, and secondly, the interval $[S_2, S_1]$. That common interval could be crossed reversibly, which runs counter to our hypotheses. Thus, the entropy can vary spontaneously only in one direction and, as we know from section 1.6.3, that direction is that of an increase.

From these results, we can extract the following conclusion:

Any spontaneous irreversible transformation within an isolated system results in an increase dS_i in the system's entropy and, if δE is the energy dissipated, we have:

$$dS_i = \frac{\delta E}{T} > 0$$

NOTE.– When a system exchanges heat with its environment, its entropy variation is:

$$dS = \frac{\delta Q}{T}$$

However, the temperature T is that of the system in question at the place where the heat exchange takes place. If, for example, the heat transfer takes place across a surface Σ and if, on Σ, the heat flux density J is not constant and nor is the temperature T, we must write:

$$\frac{dS_e}{d\tau} = \int_\Sigma \frac{J}{T} \frac{d\Sigma}{d\tau}$$

In addition to this exchange with the exterior (the index "e"), we see the superposition of an increase in internal entropy ΔS_i corresponding to the migration of heat from hot areas to less hot areas, so as to balance out the temperature and render it uniform. Ultimately, after a sufficient length of time:

$$\Delta S = \Delta S_e + \Delta S_i$$

Depending on whether the system is heated or cooled, the variation ΔS_e will either be positive or negative. However, ΔS_i is always positive.

1.6.6. *Diffusion (Fick's law)*

Consider a system in which the temperature and pressure are uniform and respectively equal to T and P. Suppose that in that system, there are two zones such that dn kilomoles move from zone A (where the chemical potential is μ_A) to zone B (where the chemical potential is μ_B).

The internal variation in entropy is such that:

$$dn T d_i S = -(\mu_B - \mu_A)dn = -d\mu dn$$

Let us involve the molar flux density N:

$$N: \quad kmol.m^{-2}.s^{-1}$$

$$N = \frac{dn}{dtdA}$$

t: time: s

A: straight area (perpendicular to the flux): m^2

$$dnT\frac{d_iS}{dt} = -NdAd\mu = -NdA\frac{d\mu}{dz}dz = -N\frac{d\mu}{dz}dV$$

z: length counted along the flux: m

Fick's law (see section 4.1.1) is written:

$$N = -\frac{D}{RT}c\frac{d\mu}{\partial z}$$

c: molar concentration: $kmol.m^{-3}$

Let us eliminate N between the last two equations:

$$dnT\frac{d_iS}{dt} = \frac{D}{RT}\left(\frac{d\mu}{dz}\right)^2 cdV$$

However:

$$\frac{dn}{c} = dV$$

Thus, locally:

$$\frac{Td_iS}{dt} = \frac{D}{RT}\left(\frac{d\mu}{dz}\right)^2 > 0$$

Once again, we see that an irreversible operation results in an increase of entropy, but it must not be forgotten that the systematic increase in entropy relates only to isolated systems (which exchange neither material nor energy with the outside world).

1.6.7. *Return of the concept of equilibrium*

In any system, whether at equilibrium or not, fluctuations of the variables of state occur, and each local fluctuation of an independent parameter causes a local decrease causes a local drop in entropy (see [PRI 99]).

$$\Delta S_i = -\frac{k_B}{2} < 0$$

The number of independent parameters is equal to the number of degrees of freedom of system, and k_B is Boltzmann's constant.

If the equilibrium is stable, the system's internal fluctuations regress spontaneously and the entropy retains the maximum value which it had left behind.

If the equilibrium is unstable, the fluctuations, which are initially microscopic, are gradually amplified until ultimately the macroscopic state is altered.

A metastable equilibrium evolves extremely slowly, giving the appearance of stability, but that equilibrium can sometimes (though not always) be shattered suddenly by a weak external force.

A system is at stressed equilibrium if the relieving of a stress causes an irreversible transformation.

A system not at equilibrium may be steady if all its *flowrates* and parameters are stable over time. For example, industrial units continuously and stably operating are in a steady state. On the other hand, a system out of equilibrium is unsteady if a single one of its parameters varies. This parameter, though, can vary either periodically or at random (which is to be avoided!).

Let us give a simple example of a system out of equilibrium but steady.

Consider 2 reservoirs R_1 and R_2 at fixed temperatures T_1 and T_2 with $T_1 > T_2$. If the heat-transfer coefficient H is constant and the surface of contact A between the two reservoirs is constant, the heat flux from R_1 to R_2 is constant.

$$\frac{dQ}{d\tau} = HA(T_1 - T_2) = \text{const.}$$

The rate of increase of the entropy of the ensemble Σ of reservoirs R_1 and R_2 is also constant, because:

$$\frac{dS}{d\tau} = \left(\frac{1}{T_2} - \frac{1}{T_1}\right)\frac{dQ}{d\tau}$$

This state is steady because the heat flux is fixed, but it is not a state of equilibrium, because the system's entropy increases. Prigogine [PRI 99] goes further and shows that, if the two reservoirs are separated by a metal bar along which the heat progresses, the rate of entropy creation in the bar is minimal if the longitudinal temperature profile in the bar is linear. Hence, this state is steady but is not a state of equilibrium.

A reversible transformation is only possible when starting from a position of stable equilibrium because, if we change the sign of the variations dx_i of the parameters, the system must return to its position initial *en route* to evolution in the opposite direction. Furthermore, if the dx_i values become 0, the initial position is regained.

1.6.8. *Thermodynamic potentials*

A thermodynamic potential X is a generator value whose first variation δX characterizes the approach to equilibrium and whose second variation $\delta^2 X$ characterizes the stability of the equilibrium. For our purposes, there are two potentials F and G.

We know that for an isolated system where a spontaneous transformation takes place:

– we approach equilibrium if $\delta S > 0$;

– equilibrium is achieved if $\delta S = 0$;

– that equilibrium is stable if $\delta^2 S < 0$ – i.e. if the entropy has reached its maximum value.

1) The Helmholtz energy F is a potential

Consider a system which exchanges heat with the outside, but not work or material, in a transformation which is not necessarily reversible. We have:

$$TdS = T(d_iS + d_eS) \geq \delta Q = dU \qquad [1.3]$$

where:

$$Td_eS = \delta Q \text{ and } d_iS \geq 0$$

Now let us introduce the Helmholtz energy:

$$F = U - TS \qquad \text{(Euler's theorem)}$$

At constant temperature and for a transformation that is not necessarily reversible:

$$TdS = dU - dF$$

Hence, if we refer to equation [1.3]:

$$dU - dF \geq dU$$

That is to say:

$$dF \leq 0$$

Thus, in these precise conditions:

$$T = \text{const.} \quad V = \text{const. (no work exchanged)}$$

The negative variation of F stems from the positive variation of S. Thus, the Helmholtz energy F is a thermodynamic potential whose decrease yields a stable equilibrium characterized by the minimum of F:

$$\delta F = 0 \quad \text{and} \quad \delta^2 F > 0$$

The stresses of constant temperature and volume could, *a priori*, exist in a thermostatic autoclave on condition that the volume of the dense phases (liquid and any solids) remain constant. Unfortunately, in these devices, there tends to be a gaseous ceiling which allows variations in volume of the dense phases.

Thus, it is much easier to keep the pressure constant than the volume; this is the advantage of the Gibbs energy, to which we now turn our attention.

2) The Gibbs energy G is a potential

Consider a system which is home to irreversible internal evolutions and which, at constant T and P, exchanges the heat with the exterior:

$$TdS = T(d_iS + d_eS) > \delta Q = dH \qquad [1.4]$$

The Gibbs energy is defined by:

$$G = H - TS \qquad \text{(Euler's theorem)}$$

and, at constant temperature:

$$TdS = dH - dG$$

If we refer to inequality [1.4] – i.e. if we eliminate TdS:

$$dH - dG \geq dH \text{ meaning that } dG \leq 0$$

Thus, the decrease in G stems from the spontaneous increase in S. The Gibbs energy G is therefore a thermodynamic potential whose decrease leads to a stable equilibrium characterized by a minimum of G:

$$\delta G = 0 \text{ and } \delta^2 G > 0$$

The stresses at constant temperature and pressure are commonly found in industry, provided the chemical reaction takes place continuously in a stirred, thermostatic vat – e.g. temperature controlled by the circulation of a working fluid (vapor, water, brine, glycolated water, etc.) in a coil and/or a double casing.

In the gaseous phase (and sometimes in the liquid or paste phase under pressure), the reaction can take place in a tubular reactor surrounded by another tube. The working fluid circulates between the two tubes and keeps the temperature constant, whilst the pressure is maintained by a compressor or a pump.

3) Conclusion

Note the similarity between the potential energy found in mechanics (available for a movement) and the Gibbs energy and Helmholtz energy (available for a chemical reaction, a transfer between two phases or transport within a phase). Therefore, the Gibbs energy G and Helmholtz energy F are potentials.

1.7. Use of the concept of entropy – thermal machines and entropy analysis

1.7.1. *Ideal performances of thermal machines*

By definition, a thermal machine operates between a hot source at the absolute temperature T_H and a cold source at the temperature T_C. The fluid in that machine runs in a closed cycle.

That fluid exchanges the heat Q_H with the hot source and exchanges the heat Q_C with the cold source. These exchanges are supposed to be reversible.

The work W received from outside is also supposed to be reversible, meaning that there is no variation in entropy.

If S is the entropy of the fluid in the machine, we can write:

$$Q_C = T_H \Delta S_H$$

$$Q_F = T_C \Delta S_C$$

Because there is no variation in entropy when the work is carried out, we can write the following about a cycle of the fluid:

$$\Delta l\, S_H + \Delta l\, S_C = 0$$

and the overall energy balance is expressed by:

$$W + Q_H + Q_C = 0$$

As usual, the *energies received by* the fluid are counted *positively*. Let us now examine a number of cases:

1) Combustion engine

$$W < 0 \qquad Q_H > 0 \qquad Q_F < 0$$

The engine's yield is:

$$\eta = \frac{-W}{Q_H} = \frac{Q_H + Q_C}{Q_H} = 1 + \frac{Q_C}{Q_H} = 1 + \frac{T_C \Delta S_C}{T_H \Delta S_H} = 1 - \frac{T_C}{T_H}$$

2) Heat pump

$$W > 0 \qquad Q_H < 0 \qquad Q_C > 0$$

$$\eta = \frac{-Q_H}{W} = \frac{-Q_H}{-Q_H - Q_C} = \frac{-T_H \Delta S_H}{-T_H \Delta S_H - T_C \Delta S_H} = \frac{T_H}{T_H - T_C}$$

Here, η is the ideal productivity of the heat pump.

3) Refrigeration unit

$$W > 0 \qquad Q_H > 0 \qquad Q_C > 0$$

$$\eta = \frac{-Q_C}{W} = \frac{-Q_C}{-Q_H - Q_C} = \frac{-T_C \Delta S_C}{-T_H \Delta S_C - T_C \Delta S_C} = \frac{T_C}{T_H - T_C}$$

Here, η is the ideal productivity of the unit.

1.7.2. *Entropy analysis*

Entropy, as such, is not widely used in process engineering. However, when the cost of a process necessitates a careful examination of its energy balance, it is essential to find those of the operations which lead to significant energy wastage.

The procedure used for this is an entropy analysis, which consists of, for each elementary operation, making *simplifying* hypotheses (e.g. assimilating a gas to a perfect gas) in order to quickly and easily estimate the energy dissipation.

Practice shows us that entropy analysis is more revealing than energy balances, where the entropy is multiplied by the ambient temperature which, often, has little to do with the process.

Let us cite a few examples of the calculation of entropy variation. As entropy analysis is generally used for installations running continuously, we shall express the entropy generation rates in Watt.K^{-1}.

1) De-stressing (approximate expression)

$$\Delta S \sim \int_0^1 W \frac{V dP}{T} \sim W \frac{\bar{V} \Delta W}{T}$$

$\overline{V}$ and $\overline{T}$ are the arithmetic means between the upstream and downstream values of the molar volume and the temperature. W is the flowrate of the de-stressed fluid in $kmol.s^{-1}$.

2) Heat exchanger (rigorous expression)

$$\Delta S = \Delta S_{cold\ fluid} + \Delta S_{hot\ fluid} = wcLn\left[\frac{t_{outputs}}{t_{inputs}}\right] + WC\ Ln\left[\frac{T_{output}}{T_{input}}\right]$$

W and w are the flowrates of the hot and cold fluids. C and c are the thermal capacities of those fluids. T and t, here, are absolute temperatures.

A reversible exchanger must have an infinite surface, but also, WC must be equal to wc. Indeed, in this case:

$$t_{output} = T_{input} \quad \text{and} \quad T_{output} = t_{input}$$

which means that $\Delta S = 0$.

3) Vaporization or condensation of a pure fluid (rigorous expression)

If Q is the thermal power of the device (in Watt), $T_{S\ input}$ and $T_{S\ output}$ the temperature of the working fluid at the input and output, and T_P the temperature of state change of the product:

$$\text{Condensation: } \Delta S = \frac{Q}{T_P} + W_S C_S Ln\frac{T_{S\ output}}{T_{S\ input}} \text{ where } Q < 0$$

$$\text{Vaporization: } \Delta S = \frac{Q}{T_P} + W_S C_S Ln\frac{T_{S\ output}}{T_{S\ input}} \text{ where } Q > 0$$

W_S and C_S are the flowrate and the specific heat capacity of the working fluid.

4) Mixture (approximate expression, with gases supposed to be perfect)

If z_i is the molar fraction of the component with index i in the mixture and, if that mixture is produced at the rate of n_i kilomoles of that component per second, and finally if it contains c components, we have:

$$\Delta s^M = -R\sum_{i=1}^{c} n_i Lnz_i \quad \text{with} \quad z_i = n_i/\sum_{i=1}^{c} n_i$$

R is the perfect gas constant (8314 J. kmol^{-1}.K^{-1}) and ΔS^M is what is called the "entropy of mixing".

5) Separation columns and gas–liquid fractional columns

The overall irreversibility of distillation columns is the sum of those of the boiler and the condenser, less the entropy of mixing of the input product.

A necessary but not sufficient condition for a column to be reversible would be that the two phases move across their equilibrium surface, meaning that the column would need to have an infinite number of platforms or that, in a packed tower, the contact surface between the phases should be infinite.

6) Generator machines (pumps, compressors, approximate expression)

We consider that, firstly, the mechanical losses of the machine (given by the no-load power) and that, secondly, the losses due to shocks and friction in the fluid accumulate. These losses are resolved in the form of heat. The heat thus generated is δQ and the entropy creation is:

$$\Delta s = \frac{\delta Q}{\overline{T}}$$

$\overline{T}$ is the arithmetic mean of the absolute temperatures of input and output of the fluid.

7) Solvent-based extraction (rigorous expression)

Let S_A, S_S, S_E and S_R represent the molar entropies of the input product, the solvent, the extract and the raffinate. Then, we would have:

$$\Delta h = W_E S_E + W_R S_R - W_A S_A - W_S S_S$$

The W values are the molar flowrates corresponding to the effluents and affluents.

8) Pressure drop in a pipe

$$\Delta s \# \frac{W \overline{V} \Delta i}{\overline{T}}$$

$\overline{V}$ and $\overline{T}$ are the arithmetic mean of the molar volume ($m^3.kmol^{-1}$) and the arithmetic mean of the absolute temperature at the two ends of the pipe. ΔP is pressure drop (in Pa). W is the flowrate in kilomoles per second.

1.8. Gibbs–Duhem and Gibbs–Helmholtz equations

1.8.1. *Gibbs–Duhem equation*

Take the total derivative of Euler's equation:

$$dU = TdS + SdT - PdV - VdP + \sum_i \mu_i dn_i + \sum_i n_i d\mu_i$$

However, we know that:

$$dU = TdS - PdV + \sum_i \mu_i dn_i$$

and, by subtracting member by member, we obtain the Gibbs–Duhem equation:

$$SdT - VdP + \sum_i n_i d\mu_i = 0$$

We shall show in section 2.4.4 that the equation $\sum_i x_i d\mu_i = 0$ results from the expression of the μ_i values.

1.8.2. *Generalized Gibbs–Duhem equation*

Consider the following extensive values:

G, S, H, U, V

Let M be the chosen value. By definition:

$$\overline{m_i} = \left[\frac{\partial(n_T m)}{\partial n_i} \right]_{P.T.n_j} \quad M = n_T m$$

$$d(n_T m) = \left(\frac{\partial(n_T m)}{\partial T} \right)_{P.n_j n_j} dT + \left(\frac{\partial(n_T m)}{\partial P} \right)_{T.n_j n_j} dP + \sum_i (\overline{m_i} dn_i)$$

or indeed:

$$d(n_T m) = n_T \left(\frac{\partial m}{\partial T}\right)_{P.n_i} dT + n_T \left(\frac{\partial m}{\partial P}\right)_{T.n_i} dP + \sum_i \overline{m_i} dn_i \qquad [1.5]$$

Yet:

$$d(n_T m) = \sum_i d(n_i \overline{m_i}) = \sum_i \overline{m_i} dn_i + \sum_i n_i d\overline{m_i} \qquad [1.6]$$

Member by member, let us subtract equation [1.6] from equation [1.5] and divide by n_T, in the knowledge that $n_i/n_T = z_i$:

$$\left(\frac{\partial m}{\partial T}\right)_{P.n_i} dT + \left(\frac{\partial m}{\partial P}\right)_{T.n_i} dP - \sum_i z_i d\overline{m_i} = 0$$

For instance, if $M = H$:

$$C_p dT + \left[V - T\left(\frac{\partial V}{\partial T}\right)_{P.z}\right] dP - \sum_i z_i d\overline{h_i} = 0$$

NOTE.– The above considerations bring into play:

– intensive variables: T, P and μ_i;

– extensive variables: U, H, F, G, V and S.

For the time being, let us give these six extensive variables the general appellation M.

We have:

$$M = \sum_{i=1}^{c} n_i \overline{m_i} \text{ where } \overline{m_i} = \left[\frac{\partial M}{\partial n_i}\right]_{T.P.n_k} (m_i \text{ is a partial value})$$

For example:

$$\overline{h_i} = \left[\frac{\partial H}{\partial n_i}\right]_{T.P.n_k} \text{ and } H = \sum_{i=1}^{c} n_i \overline{h_i}$$

It must be noted that all these partial derivations are performed with T, P and n_k constant, which distinguishes the partial values from the chemical potentials; indeed:

$$\bar{h}_i = \left[\frac{\partial H}{\partial n_i}\right]_{T.P.n_k} \neq \left[\frac{\partial H}{\partial n_i}\right]_{S.P.n} = \mu_i$$

The only exception to this rule pertains to the Gibbs energy G, for which g_i is identical to the chemical potential.

$$\bar{g}_i = \left[\frac{\partial G}{\partial n_i}\right]_{T.P.n_k} = \mu_i$$

Indeed, T, P and n_k (k between 1 and n_T) are the *natural variables* of G, because their derivatives are involved in dG.

$$dG = -SdT + VdP + \sum_{k=1}^{c} \mu_k dn_k$$

1.8.3. *Thermodynamic equations of state*

We know that:

$$U = F + TS$$

Hence:

$$\left(\frac{\partial U}{\partial V}\right)_T = \left[\frac{\partial (F+TS)}{\partial V}\right]_T = \left(\frac{\partial F}{\partial V}\right)_T + T\left(\frac{\partial S}{\partial V}\right)_T$$

However:

$$dF = -PdV - SdT = \left(\frac{\partial F}{\partial V}\right)_T dV + \left(\frac{\partial F}{\partial T}\right)_V dT$$

Immediately, we can see that:

$$\left(\frac{\partial F}{\partial V}\right)_T = -P \qquad \left(\frac{\partial F}{\partial T}\right)_V = -S$$

Let us now write the cross second derivative of F:

$$\frac{\partial^2 F}{\partial V \partial T} = -\left(\frac{\partial P}{\partial T}\right)_V = -\left(\frac{\partial S}{\partial T}\right)_T$$

Finally:

$$\left(\frac{\partial U}{\partial V}\right)_T = -P + T\left(\frac{\partial P}{\partial T}\right)_V$$

Similarly, if we work with H instead of U and with G instead of F, we obtain:

$$\left(\frac{\partial H}{\partial P}\right)_T = V - T\left(\frac{\partial V}{\partial T}\right)_P$$

1.8.4. *Gibbs–Helmholtz equation*

According to Euler's equation and the relation $H = U + PV$:

$$\frac{G}{RT} = \frac{H}{RT} - \frac{S}{R}$$

Let us find the derivative with respect to T:

$$\left(\frac{\partial\left(\frac{G}{T}\right)}{\partial T}\right)_P = \frac{1}{T}\left(\frac{\partial H}{\partial T}\right)_P - \frac{H}{T^2} - \left(\frac{\partial S}{\partial T}\right)_P$$

However:

$$\left(\frac{\partial H}{\partial T}\right)_P = C_P \quad \text{and} \quad \left(\frac{\partial S}{\partial T}\right)_P = \frac{C_P}{T}$$

Thus:

$$\frac{\partial\left(\frac{G}{T}\right)}{\partial T} = -\frac{H}{T^2}$$

By differentiating with respect to P, we obtain:

$$\left(\frac{\partial\left(\frac{G}{T}\right)}{\partial P}\right)_T = \frac{1}{T}\left(\frac{\partial H}{\partial P}\right)_T - \left(\frac{\partial S}{\partial P}\right)_T$$

However (see section 1.8.3):

$$-\left(\frac{\partial S}{\partial P}\right)_T = \left(\frac{\partial V}{\partial T}\right)_P \quad \text{and} \quad \left(\frac{\partial H}{\partial P}\right)_T = V - T\left(\frac{\partial V}{\partial T}\right)_P$$

Hence:

$$\left(\frac{\partial\left(\frac{G}{T}\right)}{\partial P}\right)_T = \frac{V}{T}$$

This gives us the Gibbs–Helmholtz equation:

$$d\left(\frac{G}{T}\right) = -\frac{H}{T^2}dT + \frac{V}{T}dP$$

and by differentiating the extensive variables with respect to n_i:

$$d\left(\frac{\mu_i}{T}\right) = -\frac{\overline{h_i}}{T^2}dT + \frac{\overline{V_i}}{T}dP$$

1.9. Entropy and statistical physics

1.9.1. *Ensembles model*

Let us measure multiple times, each time for a very short interval (10^{-8} seconds), a physical property X of a particle of material with a given mass. Note that the result of the measurements is not absolutely constant, but instead fluctuates from one measurement to the next. If the number of measurements n_T is great, we can see that n_i measurements give the value X_i of X and that, consequently, the probability of appearance of X_i is:

$$\pi_i = n_i/n_T \quad \text{(i variable between 1 and } \Omega\text{)}$$

Unfortunately, in practice, such a way of working is impossible. Therefore, a replacement solution has been devised, which is an ensemble of q_T particles (generally q_T is extremely large), divided into Ω categories, each exhibiting the property X_i and containing q_i particles where:

$$p_i = q_i/q_T \quad \text{(i variable between 1 and } \Omega\text{)}$$

The hypothesis of *ergodicity* is the assumption that, by making the correct choices, it is possible to obtain:

$$p_i = \pi_i$$

Thus, we have replaced random measurements over time by the *a priori* data of an ensemble of a copy of the system in question (which we call a particle of material). Depending on the characteristics imposed on the system, the definition of the ensemble is more or less complicated. Indeed, it is a question of defining the microscopic states which each express the value X_i of the variable X and whose total number is equal to Ω. However, these states depend on the characteristics defining the macroscopic state of the system under study.

We shall agree to say that the state of each fold is a microcanonical state.

1.9.2. *Microcanonical ensemble*

The characteristics defining the macroscopic state of the system are the given values of the following variables:

– the number n of molecules;

– the volume V;

– the energy E.

Such a system is an isolated system. The folds satisfy the same characteristics, and we make the hypothesis (which is never debunked by real-world experience) that the microcanonical states are distributed with equal probabilities. From this, we deduce that the probability of each is $1/\Omega$. On the other hand, the probability of existence of the macroscopic state is an increasing function of Ω. The most likely macroscopic state is necessarily the most stable.

As we have seen, Ω is the number of distinct microcanonical states compatible with the macroscopic state.

Let us arbitrarily divide the system into subsystems with the common index j which can, amongst themselves, exchange energy, volume and molecules. Those subsystems are supposed to be statistically independent. Therefore, the total number of microscopic states of the overall system is the product of the Ω_j:

$$- \Omega = \prod_j \Omega_j \text{ or indeed } Ln\Omega = \sum_j Ln\Omega_j$$

The comment on the *timeline* (see section 1.9.5) shows that, as time goes on, the macroscopic state of a system evolves only so that Ω increases until a maximum is reached, which always exists for systems in range of human experience (an exception to this is represented by the universe in its entirety, of which we do not know where it is heading to). Thus, this exception apart, we always have:

$$-\delta\Omega > 0 \text{ and } \delta^2\Omega < 0$$

Boltzmann's brilliant intuition was to bring into play a thermodynamic value S such that:

– its variations are less sudden than those of Ω;

– it is an increasing monotonic function of Ω;

– it is extensive like $Ln\Omega$.

Boltzmann quite simply posited:

$$S = kLn\Omega$$

k: Boltzmann's constant: 1.38062×10^{-23} J.K^{-1}.molecule^{-1}

In summary, we can state that:

"The natural (spontaneous) evolution of an isolated system always occurs in the direction of an increase in entropy until a maximum is reached".

The state corresponding to that maximum is a definitive, and therefore stable, state of equilibrium such that, as we established earlier (in section 1.6.5):

$$\delta S = 0 \qquad\qquad \delta^2 S < 0$$

1.9.3. *The canonical ensemble*

The values N and V are fixed, as is the temperature T. The energy must be calculated as a mean value around which we see fluctuations of the instantaneous energy value. Using the Lagrange multiplier method (see

[MOO 72], we can show that probability of the existence of a fold of energy E_j is:

$$p_j = \frac{g_j \exp(-\beta E_j)}{Z} \quad \text{when} \quad \beta = \frac{1}{kT}$$

where:

$$Z = \Sigma g_j \exp(-\beta E_j)$$

g_j: (discrete) number of states accessible with energy E_j

Z: partition function

Note the *similarity* between Z and the number Ω in the presentation of the microcanonical ensemble.

We can write:

$$dLnZ = \left[\frac{\partial LnZ}{\partial \beta}\right]_{E_j} d\beta + \Sigma_j \left[\frac{\partial \beta + Z}{\partial E_j}\right]_\beta dE_j \qquad [1.7]$$

However, in view of the definition of p_j:

$$\frac{\partial LnZ}{\partial E_j} = -\beta p_j$$

Hence:

$$\Sigma \frac{\partial LnZ}{\partial E_j} dE_j = E\beta \Sigma p_j dE_j$$

If we suppose, as Schrödinger did [SCH 46], that all the folds are linked to mechanical devices which exert on each fold with energy E_j the work dE_j, then the overall work received by the system will be:

$$\delta W = \Sigma_j p_j dE_j$$

Also, the internal energy is a weighted mean:

$$U = \frac{\Sigma_j E_j g_j \exp(-\beta E_j)}{Z} = -\frac{\partial LnZ}{\partial \beta}$$

Therefore, equation [1.7] becomes:

$$dLnZ = -Ud\beta - \beta\delta W$$

We can add $d(\beta U)$ to both sides of that equation. It then becomes:

$$d(LnZ + \beta U) = \beta(dU - \delta W) = \beta\delta Q$$

However, we know that:

$$\frac{\delta Q}{T} = dS \quad \text{and} \quad \beta = \frac{1}{kT}$$

Thus:

$$dS = kd\left[LnZ + \frac{U}{kT}\right]$$

Let us integrate [MOO 72]:

$$S = kLnZ + \frac{U}{T} + \text{const.} \tag{1.8}$$

Hence, using a more sophisticated model of the canonical ensemble, we can see the double nature of the entropy:

– statistical, for the first term;

– purely energetic for the second.

In reality, it is common to distinguish between three sorts of entropy:

– entropy of translation;

– "thermal" entropy (vibrations, torsions, etc.);

– entropy of configuration (arrangement, distribution, shape and size of molecules).

The first two types of entropy are energetic, whilst the third is statistical. The entropy of mixing, also, is statistical.

These results are not surprising. Indeed, in statistical physics (see [KIT 61]), it has been shown that the Helmholtz energy is written as:

$$F = -kTLnZ$$

Therefore, equation [1.8] is quite simply written:

$$U = TS + F + const.$$

According to Euler's theorem, the value of the constant is zero.

NOTE.– Melting and vaporization both result from the injection of molar latent heat (latent because the temperature does not vary during the phase change) Λ_f and Λ_v corresponding to entropy increases $\Delta S_f = \Lambda_f/T_f$ and $\Delta S_v = \Lambda_v/T_v$. However, these operations involve giving the molecules additional freedom of movement – thereby increasing the uncertainty as to their position. Thus, we can see that, here, as always, the entropy is a measure of the disorder in the system.

1.9.4. *Grand canonical ensemble*

In the grand canonical ensemble, where we replace the given value of the number N of molecules with that of their chemical potential μ, we define the grand partition function which we shall use to evaluate the properties of perfect gases. Remember that a system which exchanges energy and material with the outside is described by the grand canonical ensemble.

NOTE.– In summary of the above:

– the microcanonical ensemble supposes that n_T, V and U are constant;

– the canonical ensemble requires n_T, V and T to be constant;

– the grand canonical ensemble assumes the constancy of μ, V and T;

– the isothermal-isobaric ensemble implies the constancy of T, P and n_T.

1.9.5. *Timeline*

Suppose we have n molecules in the gaseous state, contained in a volume V_0. We can agree that these molecules do not interfere with one another, meaning that the volume v belonging to each molecule is negligible, which is tantamount to saying that each of them is assimilated to a material point and the number of positions accessible to each molecule is proportional to the volume V_0, which means that the molecules are statistically independent.

Therefore, the number of microscopic states which the system of n molecules can attain is:

$$\Omega_0 \sim V_0^n$$

If, for example, because of a closed valve, the gaseous mass occupies only a portion V_1 of the volume V_0, experience shows us that, if we open the valve, the gas immediately and spontaneously expands into the volume V_0. On the other hand, if the gas already occupies the volume V_0, it has never spontaneously come to be contained in a volume V_1 that is smaller than V_0. This fact, in the past, appeared paradoxical to certain people, who thought it sufficient to reverse the flow of time – i.e. to replace t with -t in the equations governing the mechanics of movement of the molecules.

In reality, the macroscopic gas samples which we generally observe contain a number n of molecules approximately equal to Avogadro's number – i.e. 10^{23} molecules per liter. Consequently:

$$\frac{\Omega_1}{\Omega_0} = \left[\frac{V_1}{V_0}\right]^n$$

We see that, if the gas were to release only 1/1000 of V_0 from its presence, the ratio of probabilities of spontaneous appearance of the two states would be:

$$\frac{\Omega_1}{\Omega_0} = 0.999^{10^{23}} = 0$$

Thus, we note that the passage from V_1 to V_0 is spontaneous and immediate, whereas the reverse evolution is impossible. Put differently, returning to the past is a phenomenon that remains in the realm of reverie. Time flows in an invariable direction, which brings to mind the image of an arrow, moving only in the direction indicated by its tip, with no possibility of spontaneously returning to its firing position on the archer's bow. Similarly, the entropy of an isolated system is an increasing monotonic function of time, to the detriment of the usable energies (Gibbs energy and Helmholtz energy).

1.9.6. *Onsager's reciprocity theorem*

Consider the n fluxes J_i created by the n generalized forces X_j. Experience shows that, *in the vicinity of equilibrium*, we can write:

$$J_i = \sum_{j=1}^{n} L_{ij} X_j$$

Let x_1 and x_2 be two microscopic displacements of different natures. These displacements are characterized by fluctuations, but Onsager used the principle of reversibility in the form of a mean. Let τ represent a small time interval. These means are written:

$$\overline{x_1(t)x_2(t+\tau)} = \overline{x_1(t+\tau)x_2(t)}$$

On the basis of this result, [ONS 31a, ONS 31b] showed that:

$$L_{ij} = L_{ji}$$

This is the theorem of reciprocity.

Equations of State and Fugacities

2.1. Perfect gases

2.1.1. *Definition of perfect gases*

The molecules of all gases are totally delocalized, but perfect gases are distinguished by the *absence of interactions between molecules*. Perfect gases are mainly known by their equation of state PV = nRT, but the chemical potential, entropy or Gibbs energy can only be obtained by statistical physics and its partition functions.

2.1.2. *Values of state of pure perfect gases*

In statistical physics, it is shown that the grand partition function of a perfect gas is (see [BAL 82]):

$$Z_G = \exp\left[e^{\mu/kT} V \left(\frac{2\pi mkT}{h^2} \right)^{3/2} \right] [\xi(T)]^N$$

Here, it is helpful to recap the values of certain universal constants in physics:

N : number of molecules in question;

N_A : Avogadro's number: 6.022×10^{26} molecules.kmol^{-1}

k : Boltzmann's constant: 1.380×10^{-23} J.K^{-1}.molecule^{-1}

R: perfect gas constant: 8314 J.K^{-1}.kmol^{-1}

$$R = N_A k = 6.022 \times 10^{26} \times 1.380 \times 10^{-23} = 8310 \approx 8314$$

The perfect gas constant R has a universally-accepted value equal to 8314. The same is not true of N_A and k. Thus, we have the following discord.

m: mass of a molecule: kg.molecule^{-1}

$$m = M/N_A$$

M: molar mass: kg.kmol^{-1}

h: Planck's constant: 6.62×10^{-34} J.s

μ: chemical potential: J.molecule^{-1}

The parameter ξ (T) expresses the influence of the motions internal to each molecule (rotation, stretching, etc.). As it rises, the temperature activates these different motions and, thus, the specific heat capacity of the gas increases in stages. Certain authors restrict the term "perfect gas" to monatomic gases known as "rare gases" but, for our purposes, for the sake of conciseness, "perfect gases" may be polyatomic as well, if we simply postulate the absence of intermolecular forces in these gases.

The grand potential A can be deduced from this:

$$A = -kT \, Ln \, Z_G$$

In other words:

$$A = -kT e^{\mu/kT} V \left[\frac{2\pi mkT}{h^2} \right]^{3/2} - NkT \, Ln \, \xi$$

From this, we immediately deduce (see section 10.6.1):

$$P = -\frac{\partial A}{\partial V} = kT e^{\mu/kT} \left[\frac{2\pi mkT}{h^2} \right]^{3/2} \qquad \qquad [2.1]$$

$$N = -\frac{\partial A}{\partial \mu} = e^{\mu/kT} V \left[\frac{2\pi mkT}{h^2}\right]^{3/2} \qquad [2.2]$$

This gives us the equation of state for perfect gases:

$$PV = NkT = RT(\text{si } N = N_A) \qquad [2.3]$$

In light of equation [2.1]:

$$\mu = -kT\,LnkT - \frac{3}{2}kT\,Ln\left[\frac{2\pi mkT}{h^2}\right] + kT\,Ln\,P$$

$$\mu = -\frac{5}{2}kT\,LnkT - \frac{3}{2}kT\,Ln\left[\frac{2\pi m}{h^2}\right] + kT\,Ln\,P$$

Note, in equation [2.3], that N/V is independent of m at given T and P. Thus, the molar volume of the components of a mixture of perfect gases is the same for all the components.

μ is written, for a kilomole:

$$\mu = \mu_0(T,m) + RT\,Ln\,P \quad (\text{J.kmol}^{-1})$$

We can see that the chemical potential of a perfect gas does not depend on the Helmholtz energies of excitation of the molecules. Indeed, that "potential" is the tendency of the molecules to "escape" from the phase containing them, which is expressed by the term RTLnP.

The Gibbs energy is:

$$G = N\mu = -\frac{5}{2}NkT\,Ln\,kT - \frac{3}{2}NkT\,Ln\left[\frac{2\pi m}{h^2}\right] + NkT\,Ln\,P$$

and, if N is equal to Avogadro's number N_A, the Gibbs energy is written:

$$G = -\frac{5}{2}RT\,Ln\,kT - \frac{3}{2}RT\,Ln\left[\frac{2\pi m}{h^2}\right] + RT\,Ln\,P$$

Thus, we can write:

$$G = G_0(T,m) + RT \operatorname{Ln} P$$

The entropy becomes:

$$S = -\frac{\partial A}{\partial T} = k e^{\mu/kT} V \left[\frac{2\pi mkT}{h^2} \right]^{3/2} - \frac{\mu}{T^2} T e^{\mu kT} V \left[\frac{2\pi mT}{h^2} \right]^{3/2}$$

$$+ \frac{3}{2} k e^{\mu/kT} V \left[\frac{2\pi mT}{h^2} \right]^{3/2} - Nk \frac{\partial}{\partial T}(T \operatorname{Ln} \xi)$$

and in view of equation [2.2]:

$$S = \frac{5}{2} Nk - \frac{N\mu}{T} - Nk \frac{\partial}{\partial T}(T \operatorname{Ln} \xi)$$

Let us replace μ with its value:

$$S = \frac{5}{2} Nk - Nk \left[-\operatorname{Ln} kT - \frac{3}{2} \operatorname{Ln} \left(\frac{2\pi mkT}{h^2} \right) + \operatorname{Ln} P \right] - Nk \frac{\partial}{\partial T}(T \operatorname{Ln} \xi)$$

If $N = N_A$:

$$S = \frac{5}{2} R + \frac{3R}{2} \operatorname{Ln} \left(\frac{2\pi m}{h^2} \right) + \frac{5}{2} R \operatorname{Ln} kT - R \operatorname{Ln} P - R \frac{\partial}{\partial T}(T \operatorname{Ln} \xi)$$

or, more concisely:

$$S = S_0(T,m,\xi) - R \operatorname{Ln} P$$

Here, the molecules' Helmholtz energy of excitation comes into play, because it contributes to the term ST of the Helmholtz energy.

The term in LnkT shows that the entropy of a perfect gas tends toward $-\infty$ if T tends toward zero. This runs counter to Nernst's law. Thus, the perfect gas is simply an abstraction at very low temperatures.

The enthalpy of a perfect gas, then, is:

$$H = G + TS = \frac{5}{2}RT - RT\frac{\partial}{\partial T}(TLn\xi) = H(T)$$

We note that the molar enthalpy of a perfect gas depends only on the temperature.

The Helmholtz energy is written:

$$U = H - PV = \frac{3}{2}RT - RT\frac{\partial}{\partial T}(TLn\xi) = U(T)$$

Thus, the Helmholtz energy of a perfect gas, like its enthalpy, depends only on the temperature and on its nature (by way of the ξ term).

2.1.3. *Concept of a perfect gas associated with a chemical species*

Any associated perfect gas (APG) is distinguished by its molecular mass $M = mN_A$ and by the parameter $\xi(T)$ which expresses the internal agitation in each molecule.

However, all APGs, without exception, have the same equation of state.

$$PV = RT$$

[GUI 49] gives useful elements for theoretically finding the parameter $\xi(T)$ – i.e. the specific heat capacity and therefore the enthalpy of perfect gases. In practice, though, it is common to simply refer to the broadly-used published documents (e.g. the *Chemical Engineers' Handbook* [GRE 07]) which give empirically-proven results in the form of specific heat capacities for gases (and liquids).

Hereinafter, the properties of APG will be denoted by an asterisk.

NOTE.– It is possible to precisely find the chemical potential of hydrogen at an absolute pressure of 1 bar and temperature of 20°C because, in these conditions, this gas is extremely close to a perfect gas. This gives us an approach to the potential of hydrogen half cell.

2.2. Mixture of perfect gases

2.2.1. *Preliminaries*

In this discussion, we shall determine the partial values of the different components in the mixture.

2.2.2. *Partial volume and partial pressure*

The equation of state is written:

$$P_T V_T = n_T RT \text{ (the index T means "total")}$$

The total pressure is the sum of the partial pressures:

$$P_T = \sum_i p_i^* = \sum_i y_i P_T \qquad \text{because} \qquad p_i^* = y_i P_T$$

In a perfect gas, the molecules each occupy the same partial volume v_i^* whatever the component i:

$$v_i^* = \frac{V_T}{n_T} (m^3 . \, kmol^{-1})$$

Hence:

$$p_i^* v_i^* = y_i \frac{P_T V_T}{n_T} = y_i RT$$

2.2.3. *Partial entropy*

The partial entropy is written, for the component i at pressure p_i^* and in the pure state:

$$s_i^* = s_{oi}^*(T) - RLnp_i^*$$

For the overall mixture:

$$S = \sum_i y_i s_{oi}^*(T) - R\sum_i y_i Lny_i - RLnP_T \quad \sum_i y_i = 1$$

The term on the right, Lny_i, is called the "entropy of mixing" S^M:

$$S^M = -R\sum_i y_i Lny_i > 0$$

2.2.4. Internal energy

The internal energy of a perfect gas depends only on the temperature and is extensive in nature:

$$U^* = \sum_i y_i u_i^*(T)$$

2.2.5. Enthalpy

By definition:

$$H^*(T) = U^* + PV = U^* + RT = \sum_i y_i(u_i^* + RT)$$

Finally:

$$H^*(T) = \sum_i y_i h_i^* \quad avec \quad h_i^* = u_i^* + RT$$

2.2.6. Specific heat capacities

We simply obtain, by differentiation:

$$C_P^* = dH^*/dT \quad C_V^* = dU^*/dT$$

2.2.7. Gibbs energy and chemical potential

The Gibbs energy of the component i in the pure state and at pressure p_i^* is:

$$g_i^* = h_i^*(T) - Ts_{oi}^* + RTLnp_i^*$$

For perfect gases, the Gibbs energy of the mixture is the weighted mean of the g_i^* values:

$$G^* = \sum_{i=1}^{n} y_i g_i^* = \sum_{i=1}^{n} y_i h_i^*(T) - T \sum_{i=1}^{n} y_i s_{oi}^*(T) + \sum_{i=1}^{n} y_i RTLny_i P_T$$

However,

$$\sum_{i=1}^{n} y_i h_i^*(T) = H(T), \quad \sum_{i=1}^{n} y_i s_{oi}^*(T) = S_o(T)$$

$$\sum_{i=1}^{n} y_i LnP_T = LnP_T$$

Thus:

$$G = H(T) - TS_o(T) + LnP_T + G^M$$

where:

$$G^M = \sum_{i=1}^{n} y_i RTLny_i < 0$$

We see that the operation of mixing:

– increases the entropy by S^M;

– decrease the Gibbs energy by G^M.

with:

$$TS^M + G^M = 0$$

The chemical potential μ_i^* of the component i is:

$$\mu_i^* = g_i^*$$

The S^M and G^M make the difference between, firstly, the set of unmixed components and, secondly, the correctly made-up mixture.

Indeed, the properties of a perfect gas in a mixture at temperature T and at pressure P_T are identical to those of the same perfect gas taken *in the pure state*, at the temperature T and *pressure* p_i^* with:

$$p_i^* = y_i P_T.$$

2.3. Real chemical species in the pure state

2.3.1. *Two categories of equations of state*

With a view to applications in the chemical- and petroleum industries, the most widely studied equations are of two sorts:

a) the equation of the virial which is written in two forms:

$$\frac{PV}{RT} = 1 + \frac{B}{V} \quad [OCO\ 67]$$

or indeed:

$$\frac{PV}{RT} = 1 + BP \quad \text{(set out by Renon } et\ al.\ [REM\ 71])$$

The use of this equation must be limited to moderate pressures (less than 15 bars).

Ample information about the equation of the virial is to be found in Prausnitz *et al.* [PRA 86] and, in particular, on the equivalence of the above two equations.

b) the so-called "cubic" van der Waals equations. These are of the form:

$$P = \frac{RT}{V - b} - \frac{a}{g_2}$$

g_2 is a second-degree polynomial with respect to V.

2.3.2. *Equations based on the coefficients of the virial*

1) The Benedict–Webb–Rubin equation (BWR equation)

For each pure substance, the number of parameters is considerable and the rules of mixing are complicated.

2) Bender's equation (1970)

Bender's equation requires around 20 parameters for a pure substance.

3) Plöckner's version of the Lee–Kesler equation (the LKP equation)

This equation contains 12 parameters for a pure substance. It uses elaborate laws of mixing. It is the most flexible of the equations based on the coefficients of the virial. It is valid particularly for fluids with no dipole moment whose molecules are distinguished only by their shape and size.

NOTE.– 1) For particular mixtures (distillation of air, synthetic gases: CO, CO_2, H_2, H_2O, etc.), there are high-performing equations (Bender, Benedict–Webb–Rubin, and Lee–Kesler improved by Plöckner). These equations include a significant number of parameters (sometimes more than 10) and do not appear to be generalizable in the chemical- or oil industries.

2) Nelson and Obert [NEL 54], in Figures 1 and 2 of their book, give a universal grid of curves of the compressibility coefficient Z as a function of the reduced pressure P_r. These curves are determined by the reduced temperature T_r. Remember that:

$$P_r = \frac{P}{P_c}; \text{ and } T_r = \frac{T}{T_c}; \ Z = \frac{PV}{RT}$$

P_c and T_c are the critical pressure and critical temperature of the species in question. Of course, the results to be expected from this grid of curves are not precise. Appendix 7 shows the critical pressure and temperature of the most common gases.

2.3.3. *Cubic-type equations*

Cubic equations are so called because the calculation of the molar volume as a function of the pressure and temperature requires the solving of a third-degree equation.

In 1873, van der Waals had put forward:

$$P = \frac{RT}{V-b} - \frac{a}{V^2}$$

P: pressure: Pa

V: molar volume: $m^3.kmol^{-1}$

b: covolume: $m^3.kmol^{-1}$

R: perfect gas constant: 8 314 $J.kmol^{-1}.K^{-1}$

T: absolute temperature: K

a: energy parameter: $Pa.m^6.kmol^{-2}$

In this explicit equation in pressure:

– the first term would express the perfect gas law if b were zero. In reality, the molar volume can never be zero, and becomes equal to the covolume b for an infinite pressure;

– the second term is a term of intermolecular attraction, and we see that, perfectly logically, this term is proportional to the square of the molar concentration because that concentration is equal to 1/V ($kmol.m^{-3}$).

The van der Waals equation has been the subject of the attention of numerous authors. It was improved, in particular, first Soave and then by Peng and Robinson:

$$\text{Soave's equation (1972): } P = \frac{RT}{V-b} - \frac{a}{V(V+b)}$$

Peng–Robinson's equation [PEN 76]: $P = \dfrac{RT}{V-b} - \dfrac{a}{V^2 + 2Vb - b^2}$

These two equations are practically the only ones used in the chemical- and oil industries.

In these equations, the deviation from the perfect gas law stems from two types of causes:

1) The parameter a depends on the temperature.

2) The second term on the right-hand side, as in the van der Waals equation, expresses the intermolecular attraction.

Thus, using the equation of state, we can evaluate the total differential of the internal energy:

$$dU = C_V dT + \left[T\left(\frac{\partial P}{\partial T}\right)_V - P \right] dV$$

The value in square brackets is the internal pressure, which is not zero, for the two reasons given above.

The internal pressure, as we have just seen, results from the intermolecular forces which derive from a potential:

$$F_{ij} = -\partial \Gamma / \partial r_{ij}$$

r_{ij} is the distance separating molecule i from molecule j.

The potential Γ depends, for non-spherical molecules, on the orientation of the molecule, characterized by the angles θ and ψ, with the referential axes being linked to the work space:

$$F_{ij} = -\nabla \Gamma(r_{ij}, \theta, \psi)$$

For apolar, elongated molecules, there are correlations that can be used to evaluate the coefficients a and b as a function of an empirical parameter ω

called the acentric factor. This parameter, which reports the mean influence of the angles θ and ψ, expresses the non-centric nature in the intermolecular forces. The acentric factor is essentially used for hydrocarbon chains [VID 73 pp. 55–57].

NOTE.– The term a/f(V,b), which could be called the negative pressure of attraction, makes an essential contribution to the internal pressure but is distinct from it.

Also, the cubic equations do not always very accurately express the properties of liquids. In section 2.4.16, we see a possible way (though it remains to be verified) of overcoming this failing.

2.3.4. Determination of the parameters of a cubic equation (see [EDM 61])

If the equation contained three parameters, we could envisage calculating them at the critical conditions – i.e. using the three equations:

$$P = f(V, T), \quad (\partial P / \partial T)_c = 0 \quad \text{and} \quad (\partial^2 P / \partial V^2)_c = 0$$

It is easier, though, to write that the cubic equation accepts a triple root and to identify the coefficients of the powers of V.

$$f(V, P_c, T_c) = (V - V_c)^3$$

In reality, in the immediate vicinity of the critical conditions, the cubic equations of state do not perfectly reflect the experimental results. The most prudent path, then, is to limit ourselves to two parameters a and b, and determine them by application of the least-squares method (see Nougier, [NOU 85]) to multiple isotherms, taking care to ensure that the theoretical calculation of the vapor pressure $\pi(T)$ coincides exactly with the experimental results. It is then easy to express the variations of the parameters a and b as a function of the temperature – e.g. by a 1st- or 2nd-order polynomial in T, possibly along with a term in $1/T$.

2.3.5. *Principle for calculating the properties of fluids*

The method consists of introducing a residual value M^R such that the thermodynamic variable under study, M, can be deduced from the corresponding value M^* for perfect gases:

$$M = M^R + M^*$$

To evaluate M^R, we use the fact that, for an evanescent pressure – i.e. a volume increasing indefinitely, all fluids behave like perfect gases, with the deviation becoming greater as the pressure increases from zero. Therefore, we write:

$$M^R = \int_0^P \left(dM - dM^* \right)$$

Note that the M^R values are, themselves, extensive values because they are the difference between two extensive values. We shall now calculate the residual values associated with the following extensive values:

U, H, G, F and S

All these calculations are performed at constant temperature.

NOTE.– The extensive variables are expressed in relation to a total number n of kilomoles, different to 1. This is why we multiply the perfect gas constant R by n. This formulation will be useful when we come to calculate the residual partial values.

2.3.6. *Residual internal energy and enthalpy*

We know that:

$$dU = TdS - PdV$$

Divided by dV, we obtain:

$$\frac{dU}{dV} = T\left(\frac{\partial S}{\partial V} \right)_{P,T} - P$$

However:

$$dF = -SdT - PdV \text{ and thus } \frac{\partial^2 F}{\partial T \partial V} = -\left(\frac{\partial S}{\partial V}\right)_{T,P} = -\left(\frac{\partial P}{\partial T}\right)_V$$

and:

$$dU = \left[T\left(\frac{\partial P}{\partial T}\right)_V - P\right]dV$$

This relation has already been demonstrated in section 1.8.3.

$$\frac{dU^*}{dV} = 0 \text{ because } U^* \text{ depends only on the temperature.}$$

Therefore:

$$U^R = \int_\infty^V \left[T\left(\frac{\partial P}{\partial T}\right)_V - P\right]dV \text{ and } H^R = U^R + PV - nRT$$

However, we can also proceed as follows:

$$dG = -SdT + VdP \text{ so } \frac{\partial^2 G}{\partial T \partial P} = -\left(\frac{\partial S}{\partial P}\right)_{T,V} = \left(\frac{\partial V}{\partial T}\right)_{P,T}$$

and

$$dH = TdS - VdP$$

Let us divide by dP:

$$\frac{dH}{dP} = T\left(\frac{\partial S}{\partial P}\right)_{T,V} + V = V - T\left(\frac{\partial V}{\partial T}\right)_{P,T}$$

$$H^R = \int_0^P \left[V - T\left(\frac{\partial V}{\partial T}\right)_P\right]dP \quad (H^* \text{ depends only on the temperature})$$

$$U^R = H^R - (PV - nRT)$$

EXAMPLE 2.1.–

We shall now verify that, in the calculation of the enthalpy of liquids, the influence of temperature is far greater than that of pressure.

Take the example of water.

Find the variation in enthalpy of liquid water between the state [20°C, P = 0.025 bar abs.] and the state [40°C, P = 6 bar abs.]

Data (VDI Heat Atlas)

If the pressure is very low (P = 0.025 bar abs.)

$$\text{at } 20°C \qquad V = 18.05 \times 10^{-3} \text{ m}^{3.} \text{ kmol-1}$$

$$\text{at } 40°C \qquad V = 18.15 \times 10^{-3} \text{ m}^{3.} \text{ kmol-1}$$

$$C_p = 4180 \text{ J.kg}^{-1}.K^{-1} \quad \text{so} \quad 75303 \text{ J.kmol}^{-1}.K^{-1}$$

The mean values of V, T and P are:

$$\bar{V} = \frac{10^{-3}}{2}(18.05 + 18.15) = 18.1 \times 10^{-3} \text{ m}^3.\text{kmol}^{-1}$$

$$\bar{T} = \frac{1}{2}(40 + 20) = 30°C = 303.15 \text{ K}$$

$$\bar{P} = \frac{1}{2}(6 + 0) = 3 \text{ bar abs.}$$

$$\frac{\Delta V}{\Delta T} = \frac{10^{-3}(18.15 - 18.05)}{40 - 20} = 0.540 \times 10^{-5} \text{ m}^3.\text{kmol}^{-1}.K^{-1}$$

$$\bar{T}\left(\frac{\partial V}{\partial T}\right)_P = 303.15 \times 0.540 \times 10^{-5} = 1.638 \times 10^{-3} \text{ m}^3.\text{kmol}^{-1}$$

$$dH = 75303 \times (40 - 20) + (18.1 - 1.638)10^{-3} \times 6 \times 10^5$$

$$dH = 1.506 \times 10^6 + 0.0009877 \times 10^6 = 1.506 \times 10^6 \text{ J.kmol}^{-1}$$

As is the case for most liquids, the influence of the variation in pressure is negligible in comparison to that of the variation in temperature.

2.3.7. *Residual Gibbs energy and Helmholtz energy*

We know that:

$$dG = -SdT + VdP$$

and, if T = const.

$$dG = VdP$$

In regard to the perfect gas

$$dG^* = V^* dP = \frac{nRT}{P} dP$$

$$G^R = \int_0^P VdP - \int_0^P \frac{nRT}{P} dP = \int_0^P \left(V - \frac{nRT}{P} \right) dP$$

Also, however:

$$\frac{dP}{P} = \frac{d(PV)}{PV} - \frac{dV}{V}$$

Thus:

$$G^R = \int_0^P (PV - nRT) \left(\frac{d(PV)}{PV} - \frac{dV}{V} \right)$$

$$G^R = PV - nRT - nRTLn\frac{PV}{RT} - \int_\infty^V \left(P - \frac{nRT}{V} \right) dV$$

The residual Helmholtz energy is deduced from this:

$$F^R = G^R - (PV - nRT)$$

$$F^R = \int_\infty^V \left(\frac{nRT}{V} - P \right) dV - nRTLn\frac{PV}{nRT}$$

2.3.8. *Residual entropy*

With P as the integration variable and in light of Euler's theorem:

$$S^R = \frac{H^R - G^R}{T} = \int_0^P \left[\frac{nR}{P} - \left(\frac{\partial V}{\partial T} \right)_P \right] dP + nRLn \frac{PV}{RT}$$

and with V as the integration variable:

$$S^R = \frac{U^R - F^R}{T} = \int_0^V \left[\left(\frac{\partial V}{\partial T} \right)_V - \frac{nR}{V} \right] dV + nRLn \frac{PV}{nRT}$$

2.3.9. *Fugacity coefficient*

The fugacity coefficient is defined simply by:

$$nRTLn\varphi = G^R$$

The fugacity of the fluid, then, is:

$$f = \varphi P$$

2.3.10. *Residual specific heat capacity at constant volume*

We have:

$$dU = TdS - PdV = TdS \qquad \text{because } dV = 0$$

$$TdS = \delta Q = C_V dT$$

Thus:

$$\left(\frac{dU}{dT} \right)_V = C_V \qquad\qquad C_V^R = \left(\frac{dU^R}{dT} \right)_V$$

$$C_V^R = \frac{\partial}{\partial T} \int_\infty^V \left[T \left(\frac{\partial P}{\partial T} \right)_V - P \right] dV = T \int_\infty^V \left(\frac{\partial^2 P}{\partial T^2} \right)_V dV$$

2.3.11. *Residual specific heat capacity at constant pressure*

We have

$$dH = TdS + VdP = TdS \text{ because } dP = 0$$

$$TdS = \delta Q = C_p dT$$

Hence:

$$\left(\frac{dH}{dT}\right)_P = C_P \qquad C_P^R = \left(\frac{dH^R}{dT}\right)_P$$

$$C_P^R = \frac{\partial}{\partial T}\int_0^P \left[V - T\left(\frac{\partial V}{\partial T}\right)_P \right] dP = -T\int_0^P \left(\frac{\partial^2 V}{\partial T^2}\right)_P dP$$

2.3.12. *Expression of $C_P^R - C_V^R$*

We have:

$$dH = C_P dT + \left(\frac{\partial H}{\partial P}\right)_{V,T} dP \qquad \text{or} \qquad \left(\frac{\partial H}{\partial T}\right)_V = C_P + \left(\frac{\partial H}{\partial P}\right)_{T,V}\left(\frac{\partial P}{\partial T}\right)_V$$

However:

$$\left(\frac{\partial H}{\partial T}\right)_V = C_V + V\left(\frac{\partial P}{\partial T}\right)_V$$

and:

$$\left(\frac{\partial H}{\partial P}\right)_V = V - T\left(\frac{\partial V}{\partial T}\right)_P$$

Thus:

$$C_P - C_V = T\left(\frac{\partial V}{\partial T}\right)_P\left(\frac{\partial P}{\partial T}\right)_V \quad \text{and} \quad C_P^* - C_V^* = R$$

$$C_P^R - C_V^R = T\left(\frac{\partial V}{\partial T}\right)_P\left(\frac{\partial P}{\partial T}\right)_V - R$$

Often, P is defined as a function of V and T and, as illustrated by Appendix C (closed-loop derivation):

$$\left(\frac{\partial V}{\partial T}\right)_P = -\frac{\left(\frac{\partial P}{\partial T}\right)_V}{\left(\frac{\partial P}{\partial V}\right)_T}$$

Therefore:

$$C_P - C_V = -T\frac{\left(\frac{\partial P}{\partial T}\right)_V^2}{\left(\frac{\partial P}{\partial V}\right)_T}$$

At the critical point, $(\partial P/\partial V)_T$ takes the value of 0 and, as C_V remains finite, it results that C_P increases indefinitely.

2.3.13. *Practical values*

It should be noted that, in normal conditions (1 atm and 20°C), the specific heat capacity of air is:

$$C_p \# 1000 \text{ J.kg}^{-1}.°\text{C}^{-1}$$

That of water is:

$$C_p \# 4180 \text{ J.kg}^{-1}.°\text{C}^{-1}$$

2.3.14. *Residual volume*

$$V^R = V - V^* = V - \frac{RT}{P}$$

2.3.15. *Differentials of entropy*

We have:

$$dF = \left(\frac{\partial F}{\partial T}\right)_V dT + \left(\frac{\partial F}{\partial V}\right)_T dV = -SdT - PdV$$

Hence:

$$\left(\frac{\partial S}{\partial V}\right)_T = -\frac{\partial^2 F}{\partial V \partial T} = \left(\frac{\partial P}{\partial T}\right)_V$$

Thus:

$$dS = C_V \frac{dT}{T} + \left(\frac{\partial P}{\partial T}\right)_V dV \qquad [2.4]$$

Similarly, if we use the differential of G, we would have:

$$dS = C_P \frac{dT}{T} - \left(\frac{\partial V}{\partial T}\right)_P dP \qquad [2.5]$$

2.3.16. *Isentropic compressibility modulus*

For an isentropic transformation, we deduce from equations [2.4] and [2.5]:

$$\frac{C_P}{C_V} = \gamma = -\frac{\left(\frac{\partial V}{\partial T}\right)_P dP}{\left(\frac{\partial P}{\partial T}\right)_V dV} = \left(\frac{\partial P}{\partial V}\right)_T \left(\frac{\partial V}{\partial P}\right)_S \text{ (see Appendix C) } \left|\frac{\partial V}{\partial P}\right|_S$$

The isentropic compressibility modulus can be deduced from this:

$$\frac{1}{V}\left(\frac{\partial V}{\partial P}\right)_S = \frac{\gamma}{V}\left(\frac{\partial V}{\partial P}\right)_T$$

2.3.17. *Practical evaluation of H and S as a function of T*

In general, the measurements are taken at constant pressure and the reference temperature T_0 is often taken as equal to 20°C but, in light of the critical temperature of the substance in question, this is not always possible, and sometimes we must choose a different temperature, which may be extremely low.

Through integration, we obtain:

$$H = \int_{T_0}^{T_f} C_{PC} dT + \Delta H_f + \int_{T_f}^{T_v} C_{PL} + \Delta H_v + \int_{T_v}^{T} C_{PG}^* dT$$

The meaning of the indices is as follows:

C: crystal

f: melting

L: liquid

v: vaporization

G: gas

$$S = \int_0^{T_f} C_{PC} \frac{dT}{T} + \frac{\Delta H_f}{T_f} + \int_{T_f}^{T_v} C_{PL} \frac{dT}{T} + \frac{\Delta H_v}{T_v} + \int_{T_v}^{T} C_{PG} \frac{dT}{T}$$

With regard to the entropy, we have integrated from a starting point of zero Kelvin. Indeed, the *third law of thermodynamics* is stated thus: "at absolute zero, the entropy of all the chemical species is null".

Indeed, according to Debye's theory, the specific heat capacity C_{PC} of crystals increases as T^3 in the vicinity of absolute zero.

$$S = \int_0^{T} C_{PC} \frac{dT}{T} \sim \int_0^{T} T^2 dT \rightarrow 0 \text{ if } T \rightarrow 0 \text{ K}$$

There are specific cases – quantum fluids, for example – which require special treatment, but such fluids are not of interest to us here.

NOTE.– We generally use the enthalpy to establish heat balances around a device or, more rarely, around a complete installation, and we write:

$$\sum_i H_{i\text{ affluents}} + Q_{\text{heating}} - Q_{\text{cooling}} = \sum_j H_{j\text{ effluents}}$$

In this equation, the reference temperature T_0 is eliminated and, to avoid unnecessary complications, we simply need to choose it to be outside of the domain affected by the installation. More specifically, it is usual for T_0 to be less than the lowest temperature of the procedure in question.

2.3.18. Joule–Thomson expansion (free expansion)

When a fluid circulates in a pipe, its pressure drops because of the friction it experiences with the wall of the pipe. The same is true when it passes through a regulating valve.

Pipes and taps are generally calorifuged to limit thermal losses (from hot fluids at temperature $\geq 40°C$) or frigorific losses (from fluids at $T \leq 0°C$) but also to ensure plumbers' safety (thus, we must have $-5°C < T < 60°C$)

The result of this is that the heat generated by friction is fully evacuated by the fluid, which causes an increase in its entropy which exactly compensates the effect of the drop in pressure. Thus, we can write:

$$dH = TdS + VdP = 0$$

We shall express the coefficient μ of the expansion in light of the equation of state of the fluid.

$$\mu = \left(\frac{\partial T}{\partial P}\right)_H$$

Yet:

$$dH = \left(\frac{\partial H}{\partial T}\right)_P dT + \left(\frac{\partial H}{\partial P}\right)_T dP = 0$$

so:

$$\mu = -\frac{(\partial H / \partial P)_T}{(\partial H / \partial T)_P}$$

By definition:

$$\left(\frac{\partial H}{\partial T}\right)_P = C_P \text{ (specific heat capacity at } P = \text{const.)}$$

Additionally,

$$dH = TdS + VdP$$

and:

$$\left(\frac{\partial H}{\partial P}\right)_T = T\left(\frac{\partial S}{\partial P}\right)_T + V$$

However:

$$dG = -SdT + VdP$$

Hence, by equalization of the second derivatives of G:

$$\left(\frac{\partial S}{\partial P}\right)_T = -\left(\frac{\partial V}{\partial T}\right)_P$$

and finally:

$$\mu = \frac{T(\partial V / \partial T)_P - V}{C_P}$$

2.3.19. *Use of the equation of state*

The van der Waals equation is the simplest of the so-called "cubic" equations. That equation is written:

$$(P + a / V^2)(V - b) = RT$$

When we differentiate with respect to T (at constant P):

$$-\frac{2a}{V^3}(V-b)\left(\frac{\partial V}{\partial T}\right)_P + (P + a/V^2)\left(\frac{\partial V}{\partial T}\right)_P = R$$

Thus:

$$\left(\frac{\partial V}{\partial T}\right)_P = \frac{R}{P + a/V^2 - 2a(V-b)/V^3} = R\left[P - a/V^2(1 - 2b/V)\right]^{-1}$$

Note, though, that in practice, a depends on the temperature. Hence, it can only possibly be an approximate calculation.

EXAMPLE 2.2.–

For air at 20°C and at pressure,

$a = 1.38 \times 10^5$ $R = 8314$

$b = 0.036 m^3.kmol^{-1}$ $T = 293K$

$V = 2.45 m^3.kmol^{-1}$

$$P = \frac{8314 \times 293}{2.45 - 0.036} - \frac{1.38 \times 10^5}{2.45^2} = 9.86 \times 10^5 \, Pa$$

$$\left(\frac{\partial V}{\partial T}\right)_P = \frac{8314}{9.86 \times 10^5 - (1.38 \times 10^5 / 2.45^2)(1 - 2 \times 0.036 / 2.45)}$$

$$\left(\frac{\partial V}{\partial T}\right)_P = 8.93 \times 10^{-3} m^3.K^{-1}.kmol^{-1}$$

$$\mu = \frac{293 \times 8.93 \times 10^{-3} - 2.45}{30300} = 0.55 \times 10^{-5} °C.Pa^{-1}$$

If the variation in pressure is:

$$\Delta P = -0.5 bar = -5 \times 10^4 \, Pa$$

The variation in temperature is:

$$0.55 \times 10^{-5} \times (-5 \times 10^{4}) = -0.27°C$$

2.3.20. Isothermal compressibility and elasticity moduli

1) Isothermal compressibility

$$K_V = -\frac{1}{V}\left(\frac{\partial V}{\partial P}\right)_T$$

For a perfect gas:

$$V = \frac{RT}{P} \quad \text{so} \quad K_V = -\frac{1}{V} \times \left(-\frac{RT}{P^2}\right) = \frac{1}{P}$$

2) Isothermal elasticity

$$K_P = -V\left(\frac{\partial P}{\partial V}\right)_T$$

For a perfect gas:

$$P = \frac{RT}{V} \quad \text{hence} \quad K_P = -V \times \left(-\frac{RT}{V^2}\right) = P$$

2.3.21. Coefficients of isobaric and isochoric thermal expansion

1) Isobaric expansion

$$\alpha = \frac{1}{V}\left(\frac{\partial V}{\partial T}\right)_P$$

2) Isochoric expansion

$$\beta = \frac{1}{P}\left(\frac{\partial P}{\partial T}\right)_V$$

For a perfect gas:

$$\alpha = \beta = \frac{1}{273.15} \, K^{-1}$$

2.3.22. Antoine's equation for the vapor pressure of a pure substance

This empirical equation is written:

$$\pi = \exp\left(A - \frac{B}{t+C} \right)$$

π: vapor pressure: Pa

t: temperature: °C

Three points are sufficient to determine A, B and C.

We can write:

$$Ln\frac{\pi_1}{\pi_2} = B\left[\frac{1}{t_2+C} - \frac{1}{t_1+C} \right] = \frac{B(t_1-t_2)}{(t_2+C)(t_1+C)}$$

$$Ln\frac{\pi_1}{\pi_3} = B\left[\frac{1}{t_3+C} - \frac{1}{t_1+C} \right] = \frac{B(t_1-t_3)}{(t_3+C)(t_1+C)}$$

Therefore:

$$\frac{(t_3+C)}{(t_2+C)} = \frac{(t_1-t_2)}{(t_1-t_3)} Ln\left[\frac{\pi_1}{\pi_2}\right] / Ln\left[\frac{\pi_1}{\pi_3}\right]$$

This equation gives C; one of the other two gives B and the expression of π gives A.

EXAMPLE 2.3 (Case of water).–

Index of points	1	2	3
t (°C)	100	50	20
π (Pa)	1.0135×10^5	0.12349×10^5	0.02339×10^5

$$C = 234.8 \qquad\qquad B = 4014 \qquad\qquad A = 23.51$$

$$Ln\pi = 23.51 - \frac{4014}{t + 234.8}$$

Thus, at t = 80°C:

$$\pi = \exp\left[23.51 - \frac{4014}{80 + 234.8} \right] = 47053 \text{ Pa}$$

The VDI Wärme Atlas Table gives 47390 Pa.

2.3.23. Specific heat capacity at constant pressure (practical expression)

The specific heat capacity at constant pressure can be expressed using a zero-, first-, second- or third-degree polynomial as a function of the temperature in a range which contains all the values involved in the procedure.

2.3.24. Clapeyron equation

When a liquid is at equilibrium with its vapor, the Gibbs energy of the kilomole transferred does not vary:

$$\Delta G = G_V - G_L = \Delta H - T\Delta S = 0$$

Thus:

$$\Delta S = \frac{\Delta H}{T} \qquad\qquad\qquad [2.6]$$

In addition, again at equilibrium, if we slightly vary T by the amount dT, then P varies by dP (and vice versa). Hence:

$$dG_V = V_V dP - S_V dT$$

$$dG_L = V_L dP - S_L dT$$

To express the fact that the equilibrium persists, let us subtract member by member:

$$0 = dG_V - dG_L = (V_V - V_L)dP - (S_V - S_L)dT$$

Let π represent the saturating pressure at T – that is, the equilibrium pressure:

$$\frac{d\pi}{dT} = \frac{S_V - S_L}{V_V - V_L} = \frac{\Delta S}{V_V - V_L} \qquad [2.7]$$

Let us eliminate ΔS between equations [2.6] and [2.7]:

$$\Delta H = T(V_V - V_L)\frac{d\pi}{dT}$$

This equation is useful in calculating the latent heat of vaporization, based on the vapor pressure of a pure substance, because the correlations for the calculation of that latent heat are not always reliable, and we usually have values of the vapor pressure π as a function of T. Indeed, this measurement is simpler than calorimetric study of a heat of vaporization.

EXAMPLE 2.4.–

Latent heat of water at 80°C.

At this temperature, Antoine's equation gives us:

$$\pi = 47053 \text{ Pa}$$

According to the perfect gas law:

$$V_V = \frac{8314 \times (80 + 273,15)}{47053 \times 18} = 3.4667 \text{ m}^3.\text{kg}^{-1}$$

(however, the table gives $3.407 \text{ m}^3.\text{kg}^{-1}$) and:

$$V_L \# 10^{-3} \text{ m}^3.\text{kg}^{-1}$$

(the table gives $1.029 \times 10^{-3} \text{ m}^3.\text{kg}^{-1}$).

Thus, according to Antoine's equation:

$$\Delta V = 3.467 - 0.001 = 3.466 \text{ m}^3.\text{kg}^{-1}$$

$$\frac{d\pi}{dt} = \pi \frac{B}{(t+C)^2} = \frac{47053 \times 4014}{(314.8)^2} = 1906 \text{ Pa.K}^{-1}$$

and according to the Clapeyron equation:

$$\Delta H = (80 + 273.15) \times 3.466 \times 1906 = 2332978 \text{ J.kg}^{-1}$$

The table gives $2308800 \text{ J.kg}^{-1}$.

2.3.25. Application: crossing the critical point

Let us assume that the vapor pressure of a pure substance is expressed correctly by Antoine's formula:

$$\text{Ln}\pi = A - \frac{B}{t+C}$$

With a slight error, we can write:

$$\text{Ln}\pi = A - \frac{B}{T} \text{ (where T is the absolute temperature)}$$

Differentiation gives us:

$$\frac{d\text{Ln}\pi}{dT} = \frac{B}{T^2}$$

however, Clapeyron's formula is written:

$$\frac{d\text{Ln}\pi}{dT} = \frac{\Delta H}{\pi T \Delta V}$$

By dividing member by member, we obtain:

$$\pi = \frac{T}{B} \frac{\Delta H}{\Delta V}$$

and, beyond the critical point (see section 2.3.6)

$$\pi = \frac{T}{B} \frac{dH}{dV} = \frac{T}{B} \left[T \left(\frac{\partial P}{\partial T} \right)_V - P \right]$$

This relation enables us, for example, to determine $\left(\frac{\partial P}{\partial T} \right)_V$ in the vicinity of the critical point.

2.3.26. *Vapor pressure based on a cubic*

In the critical region, the isotherms [P,V] are of three different natures.

Consider the isotherm T_1. As the pressure of the gaseous phase increases, its volume decreases, and at G, liquid begins to appear. Further compression corresponds to a decrease in volume at constant pressure – i.e. liquefaction, which is complete at point L. The pressure $\pi(T_1)$ for that liquefaction is what is commonly called the saturating vapor pressure of the fluid.

Additional compression beyond L results in slight decreases in volume corresponding to significant increases in pressure. In other words, the liquid is much less compressible than the gas.

When the chosen temperature rises, the length of the plateau $(V_G - V_L)$ decreases and, for a temperature T_2 equal to the critical temperature T_C, we note that:

$$V_G = V_L = V_C \quad \text{and} \quad P = P_C$$

V_C and P_C are the critical volume and critical pressure of the fluid.

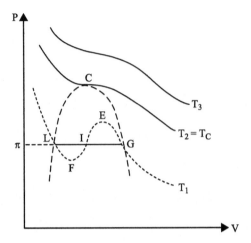

Figure 2.1. *Isotherms in the plane P–V*

In these conditions, the cubic equation of state is such that:

$$\left(\frac{\partial P}{\partial V}\right)_{T_c} = 0 \quad \text{and} \quad \left(\frac{\partial^2 P}{\partial V^2}\right)_{T_c} = 0$$

Finally, for $T_3 > T_C$, the discontinuity of the passage from the gaseous state to the liquid state no longer exists, and the more the temperature rises, the more closely the fluid's behavior resembles that of a gas.

On the arc GE, we are dealing with overcompressed gas and, on the arc LF, with overexpanded liquid but, from the point of view of the theory of equilibria, the arc LFIEG has no physical meaning, and must be replaced by the straight-line segment LG. Consider the cycle LGEFL in the form of a figure 8. For this cycle, the evolutions take place at constant temperature T_1. Thus, the variation of the Gibbs energy ΔG is written:

$$\Delta G = \oint dG = \oint VdP$$

This integral can be calculated in parts:

$$\oint VdP = [PV]_{\text{initial}}^{\text{final}} - \oint PdV$$

Because PV is a function of state, its value does not depend on the path taken, and its variation over the course of a cycle is zero. Definitively:

$$\Delta G = - \oint P dV$$

However, G is also a function of state and, over the cycle, ΔG is zero. Finally:

$$0 = \oint P dV = \int_{LI} \pi dV + \int_{IG} \pi dV + \int_{GEI} P dV + \int_{IFL} P dV$$

$$\pi(V_G - V_L) + \int_{GEIFL} P dV = 0$$

This means that the algebraic surface of the 8-cycle is null and that the areas LIFL and IEGI are equal and opposite (which is normal, as they are traversed in opposite directions).

2.3.27. *Vapor pressure and equation of state (numerical calculation)*

Our aim now is to evaluate the vapor pressure $\pi(T)$ corresponding to an isotherm at temperature $T < T_C$ using an equation of state in the form $P = f(V,T)$.

To begin with, we sweep the interval $[0,P_C]$ in successive increments equal to $P_C/30$, starting at the critical pressure P_C:

$$P_k = P_C \left(1 - \frac{k}{30} \right)$$

For each value of k, we evaluate the discriminant (see Appendix B):

$$\Delta = 4p^3 + 27q^2$$

In the vicinity of P_C (i.e. with a small value of k), Δ is positive and the cubic equation has only one root (which corresponds to a liquid).

For a value $k = \lambda$, the discriminant becomes negative and it corresponds to three roots V_L, V_I and V_G (classed in increasing order).

For a value $k = \mu$, Δ becomes positive once again and the unique root corresponds to a gas.

We divide the interval $[(\lambda - 1), \mu]$ into 30 parts and

$$P_i = P_{\lambda - 1}\left(1 - \frac{i}{30}\right)$$

For each value of P_i, we calculate V_{Li} and V_{Gi} and deduce from this the value Q_i of the surface in the form of a figure-of-8, defined in:

$$Q_i = f(V_{Li}, V_{Gi}) = f(P_i)$$

When Q_i changes sign between P_{i-1} and P_i, the desired solution is within that interval, which can, again, be divided into 30 parts. With 2 successive divisions, the precision obtained for π (T) is equal to $P_C/30^2$ – i.e. $P_C/900$.

2.3.28. *Gibbs energy of liquids*

If we want a certain degree of accuracy, it is preferable to use two different equations of state:

– for the gaseous phase, we use a form $P = f(V, T)$;

– for the liquid phase, we use the form $V = g(P, T)$.

Let $\pi(T)$ be the saturating vapor pressure of the substance in question.

1) $P < \pi$, we use $P = f(V, T)$

$$G^R_{G, \infty \to V_G} = -\int_\infty^{V_G}\left(P - \frac{RT}{V}\right) dV - RTLn\frac{\pi V_G}{RT} + \pi V_G - RT$$

We calculate the volume V_G as the largest root of the cubic equation for $P = \pi(T)$ and the chosen temperature T.

2) $P = \pi$, in which case we take $V_L = g(\pi, T)$, and we have:

$$\Delta G^R_{G \to L} = -\int_{V_G}^{V_L} \left(\pi - \frac{RT}{V} \right) dV - RTLn\frac{V_L}{V_G} + \pi(V_L - V_G)$$

Thus:

$$\Delta G^R_{G \to L} = RTLn\frac{V_L}{V_G} - RTLn\frac{V_L}{V_G} - \pi(V_L - V_G) + \pi(V_L - V_G) = 0$$

This result is normal because the liquid and gas are at equilibrium.

3) $P > \pi$, we use $V_L = g(P - \pi, T)$:

$$G^R_{L,\pi \to P} = \int_{\pi}^{P} V_L dP - RTLn\frac{P}{\pi}$$

4) In total, for the liquid at pressure $P \geq \pi$, we obtain:

$$G^R_L = -\int_{\infty}^{V_G} \left(P - \frac{RT}{V} \right) dV + \int_{\pi}^{P} V_L dP - RTLn\frac{PV_G}{RT} + \pi V_G - RT$$

We note that in studies of equilibrium, it is wrong to choose *the same* "reference state" for multiple liquids. Thus, certain authors, in the case of liquids, select a "reference pressure".

$$P_0 = 1 \text{ bar abs.}$$

and, based on the fact that the molar volume of a liquid generally varies little with the pressure (for pressures less than 30 bars), they write:

$$RTLn\phi_L = \int_{P_0}^{P} V_L dP = V_L (P - P_0)$$

However, in the expression of G^R_L, V_G depends on the nature of the fluid and, in reality, we must write (with $P = 0$ as the only "reference pressure").

$$G_L = G^*(P,T) + G^R_L$$

2.4. Properties of real mixtures

2.4.1. *Partial values (practical calculation)*

Consider an extensive value of state M. The partial values relative to the component with index i are obtained by simple differentiation:

$$\overline{m_i} = \left(\frac{\partial(n_T m)}{\partial n_i}\right)_{P,T,n_j} = \left(\frac{\partial(n_T m^*)}{\partial n_i}\right)_{P,T,n_j} + \left(\frac{\partial(n_T m^R)}{\partial n_i}\right)_{P,T,n_j}$$

This means that:

$$M = n_T m = n_T \sum_i y_i \overline{m_i} \; ; \; m = m^* + m^R \text{ and } \overline{m_i} = \overline{m_i^*} + \overline{m_i^R}$$

The values of state are the sum of the residual values and of those of the APG (marked by an asterisk). In addition, we distinguish the $\overline{m_i}$ values defined above and the m_i, which each relate to the components i in the pure state.

If we wish to calculate an energy balance (i.e. the balance either of enthalpy or entropy) around a plateau of a gas–liquid column or a slice of a liquid–liquid differential separator, we need to use the global enthalpy or entropy of the effluents and affluents.

However, if we wish to appreciate the particular influence of a component, we need to be able to evaluate the partial molar values relative to that component – particularly the partial enthalpy and partial entropy.

Finally, and this is perhaps the most important thing, the study of the equilibria between two fluids in often-complex operations (distillation, absorption, stripping, liquid-liquid extraction), requires the calculation of the chemical potentials or fugacities of each component.

2.4.2. *Fundamental nature of partial values*

We operate at constant T and P.

Consider a fluid mixture of given composition and add to it dn_i kilomoles of the component i, on the understanding that $dn_i \ll n_i$ (where n_i is the number of kilomoles of i already present in the mixture M). This means that the composition of the mixture is not modified at all, and that dn_i is a marginal variation.

By expansion into a Taylor series stopped at the first order, we obtain, by definition of $\overline{m_i}$ and for the extensive value M:

$$dM = \left(\frac{\partial M}{\partial n_i}\right)_{P,T,n_j} dn_i = \overline{m_i} dn_i$$

We can repeat this operation with all the components of the mixture. Thus, the $\overline{m_i}$ are, as we say in economic science, marginal values. They are called the "partial molar values" of components 1, 2 ..., i, ... c.

However, and this is the most interesting thing, we can reconstitute the mixture by raising the values of n_i from zero, preserving the proportions of the initial mixture and, as a consequence, we find for M,

$$M = \sum_{i=1}^{c} n_i \overline{m_i}$$

These operations of partial derivation are, in practice, used for the following values:

$$H, U, G, S, C_P, C_V$$

Furthermore, for a mixture, we have:

$$M = M^R + M^*$$

M^R is an extensive value if M is also. Thus, we have:

$$\overline{m_i} = \frac{\partial M}{\partial n_i} = \frac{\partial M^R}{\partial n_i} + m_i^* \qquad M = \sum n_i \overline{m_i}$$

We perform these derivations at constant T and P. It is preferable to know the partial enthalpies and partial entropies if we wish to correctly express the heat balances and entropy balances.

The partial Gibbs energies – i.e. the chemical potentials – are, for their part, absolutely crucial for the study of equilibria and transfers at given P and T.

2.4.3. *Practical calculation of the partial values*

Consider the extensive variable $M = n_T m$.

We calculate:

$$\overline{m}_i = \frac{\partial M}{\partial n_i} = \frac{\partial(n_T m)}{\partial n_i}$$

where:

$$n_T = \sum_{j=1}^{c} n_j; \quad m = \sum_{j=1}^{c} x_j \overline{m}_j; \quad \overline{m}_j = f(x_1,...,x_c); \quad x_j = \frac{n_j}{n_T}$$

From this we deduce (with $x_i = n_i/n_T$):

$$\frac{\partial n_T}{\partial n_i} = 1 \ \text{(i ranging from 1 to c)}$$

$$\frac{\partial x_i}{\partial n_i} = \frac{1}{n_T} - \frac{n_i}{n_T^2}\frac{\partial n_T}{\partial n_i} = \frac{1}{n_T}(1 - x_i) \ (j=i)$$

We can write:

$$\frac{\partial m}{\partial n_i} = \sum_{k=1}^{c} \frac{\partial m}{\partial x_k}\frac{\partial x_k}{\partial n_i} = \frac{1}{n_T}\left[\frac{\partial m}{\partial x_i} - \sum_{k=1}^{c} x_k \frac{\partial m}{\partial x_k}\right] \qquad [2.8]$$

Finally:

$$\overline{m}_i = \frac{\partial M}{\partial n_i} = \frac{\partial(n_T m)}{\partial n_i} = m + n_T \frac{\partial m}{\partial n_i} = m + \frac{\partial m}{\partial x_i} - \sum_{k=1}^{c} x_k \frac{\partial m}{\partial x_k}$$

NOTE.– The equations of state are generally of the form:

$$P = f(V, T, x_1, \ldots x_c)$$

In light of equation [2.8], we can always write the following, where $m = P$:

$$\left(\frac{\partial P}{\partial n_i}\right)_{V,T,n_k} = \frac{1}{n_T}\left[\frac{\partial P}{\partial x_i} - \sum_{k=1}^{c} x_k \frac{\partial P}{\partial x_k}\right]$$

However, this does not necessarily prove that P is an extensive value. Indeed (except for perfect gases):

$$P \neq \sum_{i=1}^{c} n_i \left(\frac{\partial P}{\partial n_i}\right)_{V,T}$$

2.4.4. Verification of the Gibbs–Duhem equation

Based on the molar Gibbs energy g, we can write the expression of the chemical potential:

$$\mu_i = \overline{g_i} = g + \frac{\partial g}{\partial x_i} - \sum_{k=1}^{c} x_k \frac{\partial g}{\partial x_k}$$

Let us run the calculation, bearing in mind that $\sum x_i = 1$:

$$\sum_{i=1}^{c} x_i d\mu_i = dg + \sum_{i=1}^{c} x_i d\left[\frac{\partial g}{\partial x_i}\right] - \sum_{k=1}^{c} x_k d\left[\frac{\partial g}{\partial x_k}\right] - \sum_{k=1}^{c} \frac{\partial g}{\partial x_k} dx_k$$

However:

$$dg = \sum_{k=1}^{c} \frac{\partial g}{\partial x_k} dx_k \quad (\text{with } \sum_{i=1}^{c} dx_i = 0)$$

Thus:

$$\sum_{i=1}^{c} x_i d\mu_i = 0 \ \text{(at given and constant T and P)}$$

Naturally, we can extend this result to all partial values obtained on the basis of an *extensive* value *at constant T and P*. Thus, for example:

$$\sum_{i=1}^{c} x_i d\overline{v_i} = 0 \qquad \sum_{i=1}^{c} x_i d\overline{h_i} = 0$$

Indeed, T and P appear explicitly in the expression of V and H. It must be highlighted that, for example, had we calculated $(\partial H / \partial n_i)$ at constant P and S, we would have obtained not the partial value h_i, but the chemical potential μ_i, but such a calculation is never performed because of its complexity and, to obtain μ_i, it is easier to derive g directly.

2.4.5. *Partial molar volumes based on an equation of state P = P (V,T)*

These volumes $\overline{v_i}$ are defined by:

$$\sum_{i=1}^{c} n_i \overline{v_i} = V$$

Put differently:

$$\overline{v_i} = \left(\frac{\partial V}{\partial n_i} \right)_{P,T,n_j}$$

Suppose we have an equation of state, and choose V and the number n_i of kilomoles the species i as independent variables. For $j \neq i$, the n_j are constant, as is the temperature T. The pressure varies in accordance with V and n_i.

$$dP = \left(\frac{\partial P}{\partial V} \right)_{n_i,n_j,T} dV + \left(\frac{\partial P}{\partial n_i} \right)_{V,T,n_j} dn_i$$

Divide this equation by $(\partial P / \partial V)_{n_i,T}$; we obtain (see Appendix 3):

$$dV = \frac{1}{\left(\dfrac{\partial P}{\partial V}\right)_{n_i,n_j,T}}dP - \frac{\left(\dfrac{\partial P}{\partial n_i}\right)_{V,T,n_j}}{\left(\dfrac{\partial P}{\partial V}\right)_{n_i,n_j,T}}dn_i = \left(\frac{\partial V}{\partial P}\right)_{n_i,n_j,T}dP + \left(\frac{\partial V}{\partial n_i}\right)_{P,T,n_j}dn_i$$

However (see Appendix 3):

$$\left(\frac{\partial V}{\partial P}\right)_{n_i,n_j,T} = 1 \Big/ \left(\frac{\partial P}{\partial V}\right)_{n_i,n_j,T}$$

Consequently:

$$\overline{V}_1 = \left(\frac{\partial V}{\partial n_i}\right)_{P,T,nj} = -\left(\frac{\partial P}{\partial n_i}\right)_{V,T,nj} \times \left[\left(\frac{\partial P}{\partial V}\right)_{ni,nj,T}\right]^{-1} \tag{2.9}$$

There is a simple way of calculating the partial volumes on the basis of an equation of state, in the form:

$$P = P(T, V, n_1, \ldots, n_i \ldots, n_c)$$

2.4.6. *Two-derivatives theorem*

If P and V are the only variables, we have:

$$dV - \left(\frac{\partial V}{\partial P}\right)_{T,n_i,n_j}dP = 0$$

Multiply by $\left(\dfrac{\partial P}{\partial n_i}\right)_{V,T,n_j}$ and use the expression of $\overline{V}_i$:

$$\overline{V}_i = -\left(\frac{\partial P}{\partial n_i}\right)_{V,T,n_j} \times \left(\frac{\partial V}{\partial P}\right)_{T,n_i,n_j} = \left(\frac{\partial V}{\partial n_i}\right)_{P,T,n_j}$$

If P and V are the only variables and are functions of n_i, we obtain:

$$\left(\frac{\partial P}{\partial n_i}\right)_{V,T,n_j} dV + \left(\frac{\partial V}{\partial n_i}\right)_{P,T,n_j} dP = 0$$

As the two derivatives are always positive, we can see that V and P always vary *in opposite directions*. By multiplying by n_i and finding the sum of the indices i, we obtain:

$$\sum_i n_i \left(\frac{\partial P}{\partial n_i}\right)_{V,T,n_j} dV + VdP = 0 \text{ (for } T = \text{const.)}$$

where:

$$\sum_i n_i \left(\frac{\partial P}{\partial n_i}\right)_{V,T,n_j} \neq P \quad \text{and} \quad \sum_i n_i \left(\frac{\partial V}{\partial n_i}\right)_{P,T,n_j} = V$$

Indeed, the pressure P is not an extensive variable.

2.4.7. *Rules of derivation of the integrals*

If M is the value of the overall mixture, we need to obtain $\bar{m}_i = \partial M/\partial n_i$. However, the calculation includes the derivation of definite integrals. With regard to this, let us recap some mathematical results.

Suppose we wish to differentiate, with respect to a variable n, the definite integral:

$$I = \int_{x_1}^{x_2} X dx$$

We suppose that the integrand X and the integration limits x_1 and x_2 depend on n. We can then show that:

$$\frac{\partial I}{\partial n} = \int_{x_1}^{x_2} \frac{\partial X}{\partial n} dx + \left[X \frac{dx}{dn}\right]_{x_1}^{x_2} \qquad [2.10]$$

If the integration limits do not depend on n, we simply have:

$$\frac{\partial I}{\partial n} = \int_{x_1}^{x_2} \frac{\partial X}{\partial n} dx \qquad [2.11]$$

In practice, the residual partial values are calculated at given, fixed T and P, which are generally the working conditions for the fluid in question. This is why we shall see, on the example of the partial Gibbs energies that if the integration variable is the pressure, we need to use equation [2.11], whereas if the integration variable is the volume, then it is equation [2.10] that we need to use.

2.4.8. *Correspondences between the activity coefficient and fugacity coefficients*

Consider a liquid phase *at equilibrium* with a gaseous phase. Each phase is a mixture. The components of the liquid are liquids whose vapor pressure is not equal to zero. With the fugacities, we can write, for a component i and at equilibrium:

$$\hat{f}_i^L = \hat{f}_i^G = \hat{f}_i$$

$\hat{f}_i^L$ and $\hat{f}_i^G$: fugacities of the component i in the liquid and in the gaseous mixture: Pa.

The capped characters always correspond to the mixture.

Now consider the fugacities f_i^L and f_i^G of the component i at the total pressure P *in the pure state*. We write:

By definition of the fugacity coefficients:

$$\hat{f}_i = y_i \varphi_i^G P = \gamma_i x_i \varphi_i^L P = \hat{\varphi}_i^L P$$

γ_i: activity coefficient of component i in the liquid phase.

and, even more simply, *which is a commonly-used formulation* (see section 3.1.2):

$$y_i P = \gamma_i x_i \pi_i(T)$$

π_i (T): vapor pressure of component i in the pure state: Pa.

This is tantamount to saying that:

$$\varphi_i^G = 1, \text{ that } \varphi_i^L P = \pi_i(T) \text{ and that } \hat{\varphi}_i^L = \gamma_i x_i \varphi_i^L = \hat{\varphi}_i^G$$

NOTE.– By definition, in a real mixture:

$$\mu_i = \mu_i^\circ(T) + RT\ln\hat{f}_i$$

and:

$$\hat{f}_i = \hat{\varphi}_i P \text{ where } RT\ln\hat{\varphi}_i = \left(\frac{\partial G}{\partial n_i}\right)_{P,T,n_j}$$

2.4.9. Properties of the fugacity coefficient in a mixture

In a mixture of perfect gases of the same composition, we have:

$$g_i^* = nRTLny_i P + g_{oi}^*(m_i, T)$$

Thus, by definition of φ_i, the coefficient of the component i in the mixture:

$$\overline{g}_i^R = \overline{g}_i - g_i^* = RTLn\frac{\hat{f}_i}{y_i P} = RTLn\hat{\varphi}_i$$

For the mixture as a whole, we define a fugacity f and a fugacity coefficient φ by:

$$G^R = \sum_i n_i \overline{g}_i^R = ng^R = nRTLn\frac{f}{P} = nRTLn\varphi$$

However, as G^R is an extensive value:

$$G^R = \sum_i n_i g_i^R$$

Put differently:

$$nLn\varphi = \sum_i n_i Ln\widehat{\varphi}_i$$

2.4.10. *Partial residual Gibbs energy*

By definition:

$$G^R = G - G^*$$

However, we know that:

$$dG = VdP - SdT + \sum \mu_i dn_i$$

$$dG^* = V^* dP - S^* dT + \sum \mu_i^* dn_i$$

Therefore:

$$\left(\frac{\partial G}{\partial P}\right)_{T,n_i,n_j} = V \quad \text{and} \quad \left(\frac{\partial G^*}{\partial P}\right)_{T,n_i,n_j} = V^*$$

$$\frac{\partial G^R}{\partial P} = \frac{\partial G}{\partial P} - \frac{\partial G^*}{\partial P} = V - V^*$$

Let us differentiate with respect to n_i:

$$\frac{\partial^2 G^R}{\partial P \partial n_i} = \frac{\partial V}{\partial n_i} - \frac{\partial V^*}{\partial n_i}$$

and then integrate with respect to P:

$$\left(\frac{\partial G^R}{\partial n_i}\right)_{T,P,n_j} = \int_0^P \frac{\partial^2 G^R}{\partial P \partial n_i} dP = \int_0^P \left(\frac{\partial V}{\partial n_i} - \frac{\partial V^*}{\partial n_i}\right) dP$$

Hence:

$$\left(\frac{\partial G^R}{\partial n_i}\right)_{P,T,nj} = \int_0^P \left[\left(\frac{\partial V}{\partial n_i}\right)_{P,T,nj} - \frac{RT}{P}\right] dP \qquad [2.12]$$

Indeed:

$$\left(\frac{\partial V^*}{\partial n_i}\right)_{P,T,n_j} = \frac{RT}{P}$$

In addition, we know that, for a pure substance or a mixture taken as a whole:

$$G^R = PV - nRT - nRTLn\frac{PV}{nRT} - \int_\infty^V \left(P - \frac{RT}{V}\right) dV$$

We shall show that this expression enables us to work back to expression [2.12].

According to the law of derivation of integrals:

$$\left(\frac{\partial G^R}{\partial n_i}\right)_{P,T,n_j} = P\overline{V_i} - RT - RTLn\frac{PV}{nRT} - nRT\left(\frac{\overline{V_i}}{V} - \frac{1}{n}\right) -$$

$$\int_\infty^V \left(\frac{\partial P}{\partial n_i} - \frac{RT}{V}\right) dV - \left[P - \frac{nRT}{V}\right]\overline{V_i}$$

(Indeed, the term in square brackets is 0 when $V \to \infty$)

$$\left(\frac{\partial G^R}{\partial n_i}\right)_{P,T,nj} = -RT[LnPV]_{\text{perfect}}^{\text{real}} + [RTLnV]_\infty^V - \int_\infty^V \frac{\partial P}{\partial n_i} dV \qquad [2.13]$$

Let us use the two-derivatives theorem:

$$\left(\frac{\partial G^R}{\partial n_i}\right)_{P,T,n_j} = -\left[RTLnP\right] + \int_\infty^P \frac{\partial V}{\partial n_i} dP = \int_0^P \left(\frac{\partial V}{\partial n_i} - \frac{RT}{P}\right) dP$$

We find equation [2.12] again.

Furthermore, equation [2.13] is written:

$$\left(\frac{\partial G^R}{\partial n_i}\right)_{P,T,nj} = -RTLn\frac{PV}{RT} - \int_\infty^V \left(\frac{\partial P}{\partial n_i} - \frac{RT}{V}\right)dV \qquad [2.14]$$

We see that we can use:

– the formula [2.12] if P is the integration variable;

– the formula [2.14] if V is the integration variable.

2.4.11. *Other residual partial values*

In light of equation [2.14]:

$$\overline{g}_i^R = -RTLn\frac{PV}{RT} - \int_\infty^V \left[\left(\frac{\partial P}{\partial n_i}\right)_{V,T,n_j} - \frac{RT}{V}\right]dV$$

$$\overline{u}_i^R = \left(\frac{\partial U^R}{\partial n_i}\right)_{P,T,n_j} = \int_\infty^V \left[T\left(\frac{\partial^2 P}{\partial T\partial n_i}\right)_{n_j} - \left(\frac{\partial P}{\partial n_i}\right)_{V,T,n_j}\right]dV + \overline{v}_i T\left(\frac{\partial P}{\partial T}\right)_{V,n_i,n_j} - \overline{v}_i P$$

$$\overline{h}_i^R = \left(\frac{\partial H^R}{\partial n_i}\right)_{P,T,n_j} = P\overline{v}_i - RT + \left(\frac{\partial U^R}{\partial n_i}\right)_{P,T,n_j}$$

$$\overline{h}_i^R = \int_\infty^V \left[\left(\frac{\partial^2 P}{\partial T\partial n_i}\right)_{n_j V} - \left(\frac{\partial P}{\partial n_i}\right)_{V,T,n_j}\right]dV + \overline{v}_i T\left(\frac{\partial P}{\partial T}\right)_{V,n_i,n_j} - RT$$

$$\overline{s}_i^R = \frac{\overline{h}_i^R - \overline{g}_i^R}{T} = \int_\infty^V \left[\frac{\partial^2 P}{\partial T\partial n_i} - \frac{R}{V}\right]dV + RLn\frac{PV}{RT} + \overline{v}_i\left(\frac{\partial P}{\partial T}\right)_{V,n_i,n_j} - R$$

2.4.12. *Cubic equation of state: calculation of φ*

The equation of state is written:

$$P = \frac{RT}{V-b} - \frac{a}{g_2} \quad \text{where} \quad g_2 = (V-V_1)\ (V-V_2)$$

and

$$\frac{1}{g_2} = \frac{1}{V_1-V_2}\left(\frac{1}{V-V_1} - \frac{1}{V-V_2}\right)$$

Thus:

$$-\int_\infty^V (PV-RT)\frac{dV}{V} = -\int_\infty^V\left[\frac{RT}{V-b} - \frac{a}{g_2}\right]dV + RT\int_\infty^V \frac{dV}{V}$$

and:

$$RT\ \text{Ln}\varphi = PV - RT - RT\text{Ln}\frac{PV}{RT} - \left[RT\text{Ln}\frac{V-b}{V} - \frac{a}{V_1-V_2}\text{Ln}\left(\frac{V-V_1}{V-V_2}\right)\right]_\infty^V$$

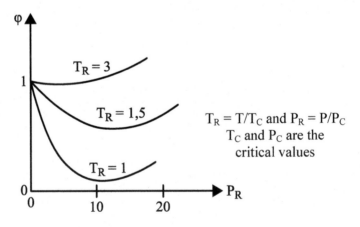

Figure 2.2. *Shape of variations of the fugacity coefficient of a gas*

NOTE.– Appendices A, B and C set out mathematical principles and tools that are useful for *practical calculations*. Remember that any integration must begin with the breaking down of the rational fractions into simple elements.

2.4.13. *Peng–Robinson equation and Helmholtz energy*

Like any cubic equation of state, the Peng–Robinson equation is written:

$$P = \frac{RT}{V-b} - \frac{a}{(V_1 - V_2)}\left[\frac{1}{V-V_1} - \frac{1}{V-V_2}\right]$$

where:

$$V_1 = b\left(\sqrt{2}-1\right) > 0 \qquad V_2 = -b\left(\sqrt{2}+1\right) < 0$$

Note that the roots V_1 and V_2 are never achieved, because always, $V > b$.

As we know from section 2.3.7, the residual Helmholtz energy is:

$$F^R = -RTLn\frac{P(V-b)}{RT} - \frac{a}{2\sqrt{2}b}Ln\left[\frac{V-b\left(\sqrt{2}-1\right)}{V+b\left(\sqrt{2}+1\right)}\right]$$

2.4.14. *Wong and Sandler's first rule of mixing [WON 92]*

When the *pressure increases indefinitely*, i.e. when V approaches b:

– the product $P(V-b)$ tends toward RT (in accordance with the equation of state);

– the logarithm becomes:

$$Ln\left(\frac{2-\sqrt{2}}{2+\sqrt{2}}\right) = Ln\left(\frac{\sqrt{2}-1}{\sqrt{2}+1}\right) = Ln\left[\frac{\left(\sqrt{2}-1\right)^2}{2-1}\right] = 2Ln\left(\sqrt{2}-1\right)$$

Therefore:

$$F_\infty^R = C\frac{a}{b} \quad \text{where} \quad C = \frac{1}{\sqrt{2}}\text{Ln}\left(\sqrt{2}-1\right)$$

and the authors write:

$$F_\infty^R - \sum_i x_i F_{\infty i}^R = C\left(\frac{a}{b} - \sum_i x_i \frac{a_i}{b_i}\right) = F_\infty^E = RT\sum_i x_i \frac{\sum_j x_j G_{j,i} \tau_{j,i}}{\sum_k x_k G_{k,i}}$$

It so happens that this expression was used by Renon with a different meaning, in the study of liquid–liquid equilibria. It has the advantage of being highly flexible. We have:

$$G_{j,i} \neq G_{i,j} \qquad G_{i,i} = 1 \qquad \tau_{i,j} \neq \tau_{j,i} \qquad \tau_{i,i} = 0$$

where:

$$G_{i,j} = \exp\left(-\alpha_{i,j}\tau_{i,j}\right) \qquad \tau_{i,j} = \frac{C_{i,j}}{RT} \qquad \alpha_{i,j} = \alpha_{j,i}$$

2.4.15. *Wong and Sandler's second rule of mixing [WON 92]*

When $V \geq 10\,b$ (the equation already corresponds to a pressure of around 100 bars), we can write:

$$\frac{1}{V-b} = \frac{1}{V}\left(\frac{1}{1-b/V}\right) \sim \frac{1+b/V}{V}$$

The cubic becomes:

$$\frac{P}{RT} = \frac{1+b/V}{V} - \frac{a/RT}{g_2(V)} \Rightarrow \frac{PV}{RT} = 1 + b/V - \frac{a/RT}{g_2/V}$$

$g_2(V)$: 2nd-degree polynomial in V.

If we have (Peng–Robinson equation):

$$g_2 = V^2 \left(1 + \frac{2b}{V} - \frac{b^2}{V^2} \right) \# V^2$$

then we obtain:

$$\frac{PV}{RT} = 1 + \frac{B}{V}, \text{ where } B = b - a / RT$$

B is the "second coefficient of the virial". Indeed, the equation of the virial is written:

$$\frac{PV}{RT} = 1 + \frac{B}{V} + \frac{C}{V^2} \text{ etc. } \ldots$$

With regard to mixtures, statistical physics shows us that the coefficient B is a quadratic function of the molar fractions:

$$B = \sum_{i=1}^{n} \sum_{j=1}^{n} z_i z_j B_{i,j}$$

Hence, the parameters a and b of the mixture must be such that:

$$b - \frac{a}{RT} = \sum_i \sum_j z_i z_j \left(b - \frac{a}{RT} \right)_{i,j}$$

This mixing law tends to be *rigorously true* when V increases indefinitely, meaning when the *pressure tends toward zero*.

We set:

$$\left(b - \frac{a}{RT} \right)_{i,j} = \frac{\left(b_i - \dfrac{a_i}{RT} \right) + \left(b_j - \dfrac{a_j}{RT} \right)}{2} \left(1 - k_{i,j} \right)$$

CONCLUSION.–

These rules govern the behavior of the mixture when:

one for $P \to \infty$ and the other for $P \to 0$

It should come as no surprise that good results have been obtained in the calculation of liquid–vapor equilibria involving polar molecules (in particular, associated liquids such as water and alcohols) and systems under pressure.

However, all cubic equations have a weakness: they are inaccurate when it comes to the molar volume of the liquid. We shall now examine a proposed strategy to overcome this weakness.

2.4.16. *An equation of state that is probably better suited to liquids*

As V approaches the covolume b, the "cubic" equations do not correctly express the liquid state. However, the molecular dynamics means that we can write, *in the absence* of forces of attraction:

$$P = \frac{RT}{V}\left[\frac{1+\xi+\xi^2-\xi^3}{(1-\xi)^3}\right] \text{ where } \xi = b/V < 1$$

This rational fraction can be decomposed into simple elements and, taking account of the term of attraction, the expression of the pressure would finally be:

$$P = \frac{RT}{V} + P_R - P_A$$

where, respectively, for the terms of repulsion and attraction, we have:

$$P_R = RT \frac{2b}{(V-b)^2}\left[2+\frac{b}{V-b}\right] \text{ and } P_A = \frac{a}{V^2+cV}$$

However, to date, this type of equation of state has not become widely used by practitioners. In addition, in the areas of distillation, absorption, stripping and liquid–liquid extraction, what is important is not the exact value of the molar volumes but the composition of the phases present, and those compositions depend only slightly on the molar volume of the liquid. As regards the molar volume of the gaseous phase, it is correct as given by the cubic equations.

2.4.17. *Calculation of the critical points*

In industrial practice, it is rare to calculate the critical temperatures and pressures of the mixtures. Thus, we shall limit ourselves to mentioning the work of Peng *et al.* [PEN 77], Heidemann *et al.* [HEI 80], Michelsen [MIC 80] and Michelsen *et al.* [MIC 81].

2.5. Characterization of crude-oil-based mixtures

2.5.1. *ASTM boiling point curve*

The term "distillation curve" is a misnomer, because the procedure merely consists of boiling the mixture in question in a flask and noting the change in temperature of the liquid mixture as a function of the vaporized fraction. The temperature of the liquid in the flask is identical to that of the vapor escaping from it.

This measurement is easy to perform. The same cannot be said of the "true boiling point" (TBP) curve, which we shall now describe.

2.5.2. *TBP curve*

The TBP was developed by American oil engineers.

The TBP curve represents the variations of a certain temperature T_v as a function of the vaporized fraction (expressed in volume of liquid) – in other words, the degree of vaporization.

The device is a flask with a column on top (such as an Oldershaw column, but extremely high), equivalent to dozens of theoretical plates, and operating in the vicinity of total reflux – i.e. with a rate of reflux of around 15 to 30. The apparatus is described in Wuithier ([WUI 65] Vol. I, page 57). The procedure is performed at atmospheric pressure.

The temperature T_v measured is the temperature *at the head* of the column. Because of the near-perfect rectification, at all times, the distillate is a practically-pure chemical species, whose boiling- and dew points coincide with the temperature T_v.

In fact, for theoretical calculations, it is necessary to use a molar rate of vaporization rather than a rate expressed in volume. We then obtain what could be called the *rational TBP curve*. Wuithier (Vol. I, p. 51-53) gives a method for evaluating the mean molar mass of a crude oil fraction. This method requires a measured value of the density d_{15} (at 15°C) of the fraction in question.

For a mixture of gasolines, the TBP curve exhibits half a dozen successive plateaus. On the other hand, for heavier fractions, the platforms are less clear. In fact, crude oils are a mixture of around 200 components which can only be separated by chromatography.

For heavy products, Edmister [EDM 61] proposed a theory of distillation using volatility (which we will see later on) varying continuously.

2.5.3. *Properties of oil fractions*

The methods for directly calculating the number of plates or for simulating an existing column were developed for discrete mixtures; not for continuous mixtures.

It is therefore necessary to divide the input into a number of fractions equal to the number of plates on the TBP curve. The successive measurements carried out when establishing the rational TBP curve immediately give us:

– the d_{15} of the elementary fractions (density at 15°C);

– the boiling point of those fractions at a pressure of 1 atm.

To determine the complete curve of the vapor pressure, it is possible to use the volatility method. In order to do so, we choose a reference substance (boiling, for example, at 175°C at 1 atm). For that substance:

$$\text{Ln}\pi_o(T) = A_o - \frac{B_o}{T}$$

Any given substance is therefore characterized by its volatility α – that is to say:

$$\pi = \alpha\pi_0 \qquad \text{meaning that} \qquad Ln\pi = Ln\alpha + Ln\pi_0$$

If the volatility were independent of the temperature, in the universal representation, all the lines of vapor pressure $Ln\pi = f(1/T)$ would be parallel to one another, because only the parameter A would change from one substance to another:

$$A = A_0 + Ln\alpha$$

Yet the reality is different, and that is why α must vary not only with the boiling point T_{boil} of the substance at atmospheric pressure, but also with the temperature T of the system.

$$\alpha = \alpha(T_{boil}, T)$$

A graph showing the shape of this function is given in Edmister [EDM 61, p. 162]. However, that graph is presented as being approximate.

To simulate the distillation in a vacuum of the atmospheric residue, we can work in the same way but, this time, use the TBP curve at 10 torr converted to the pressure of the column in a vacuum.

Pressurized distillations, for their part, relate to volatile, well-known chemical species (butane, pentane, isobutane, etc.), and thus for each substance, we simply need to adjust the parameters of a cubic equation of state which then tells us the necessary properties.

Thus, each fraction can be characterized by:

– its density d_{15};

– its mean molar mass;

– its saturating vapor pressure $\pi(T)$;

– its latent heat, obtained using the Clapeyron equation;

– its specific heat capacity (see [WUI 65], Vol. I, pp. 86–87, or [KER 50], pp. 806–807).

If these properties are known, it becomes possible to run the calculations for liquid–vapor equilibrium for the distillation, using the Peng–Robinson equation in combination with Wong and Sandler's rules of mixing where, generally:

$$F^E \equiv 0$$

2.6. Conclusion

2.6.1. *Advantage to equations of state*

The equations of state are absolutely essential in determining the enthalpy and entropy of a pure substance, which we need in order to establish its Mollier diagram, which is very widely used in the engineering of machines such as compressors and turbines. In addition, precise calculations of entropy are needed if we want to carry out an entropy analysis of an installation (see section 1.7.2). Finally, the fugacity coefficients are commonly used in the petroleum industry.

2.6.2. *A very useful simplification*

In the calculations pertaining to transfer between two fluids, the greatest possible care must be taken in order to calculate the fugacities and the activities – i.e. the corresponding coefficients φ_i and γ_i. However, it is possible to make simplifications, and appropriate results are obtained in the fields of distillation, absorption and stripping, using simplified expressions for the partial enthalpies:

Liquids $\overline{h_i^L} = h_{io}^L + C_{Pi}^L(T - T_o)$

Vapors $H_i^V = h_{io}^V + \Lambda_{io} + C_{Pi}^V(T - T_o)$

Λ_{io}: heat of vaporization of the component i at T_o and P: $J.kmol^{-1}$.

h_{io}^L: enthalpy of component i in the pure liquid state at T_o and P. However, as the calculations only involve differences in enthalpy, it is possible to take $h_{io} = 0$, provided that for T_o, we choose a value less than all the temperatures involved in the calculations.

C_{Pi}: molar specific heat capacities of the component i in the pure liquid state and in the pure vapor state: $J.K^{-1}.kmol^{-1}$.

Solution Activity Coefficients

3.1. Mixtures in the liquid state (non-electrolytes)

3.1.1. *Raoult's law for ideal solutions*

Raoult considered that, at pressure P, at temperature T and for the molar fractions y_i and x_i of the component i respectively in the vapor and in the liquid at equilibrium with the vapor:

– the vapor phase is a perfect gas in which the fugacity of the component i is y_iP;

– in the liquid phase, the fugacity of the component i is $x_i\pi_i(T)$, where $\pi_i(T)$ is the vapor pressure of the substance i in the pure state.

Equilibrium is expressed by the equality of the fugacities – i.e., according to Raoult's law:

$$y_iP = x_i\pi_i(T)$$

3.1.2. *How to describe real solutions*

There are only a limited number of solutions whose behavior corresponds to that of ideal solutions. Therefore, a way of supplementing Raoult's law has had to be found.

With this in mind, in the expression of the fugacity, f_i, we introduce an additional coefficient γ_i such that:

$$\hat{f}_i^L = \gamma_i x_i \pi_i$$

In particular, we can define the activity a_i by:

$$a_i = \gamma_i x_i \qquad \text{so} \qquad \hat{f}_i^L = a_i \pi_i$$

The coefficient γ_i is the activity coefficient, and comparison with reality shows that, generally, γ_i varies *only slightly with changing temperature* and greatly with the composition of the solution. Thus, very often, it is possible to represent liquid–vapor equilibria by completing ignoring the variation of the γ_i values as a function of the temperature. Finally, the activity coefficients can be considered to be totally independent of the pressure. If this were not the case, we would need to abandon solutions theory and use functions of state $P = P\ (V,\ T,\ \text{composition})$. Also, practice shows that it is often acceptable to consider the gaseous phase to be perfect. Ultimately, we shall use the term "*simple equilibrium equation*" to speak of the equation:

$$y_i P = \gamma_i x_i\ \pi_i$$

3.1.3. *Validity of solutions theory*

In principle, this theory applies only to solutions where *all* of the components have a critical temperature greater than the temperature of the solution (subcritical components). If supercritical components are involved, we need to use equations of state $P = f\ (V,\ T)$ or indeed $V = g\ (P,T)$, with rules of mixing. However, if a component is supercritical, we can often obtain reasonable results by extending its vapor pressure curve beyond its critical point.

3.1.4. *Excess Gibbs energy for the liquid phase*

In real solutions, the intermolecular forces are no longer independent of the substances present. To express this fact, we bring into play the excess energy, which is linked to the potentials of intermolecular action.

Specifically, this excess energy is the excess Gibbs energy G^E, for which we shall give the expression later on. The advantage of this energy is that it can be used to directly calculate the excess values μ_i^E and γ_i.

From $G^E = G^E (T, n_1, \ldots, n_i \ldots n_c)$, we deduce:

$$\mu_i^E = RTLn\gamma_i = \left(\frac{\partial G^E}{\partial n_i}\right)_{T,n_j}$$

$$s_i^E = \left(\frac{\partial^2 G^E}{\partial n_i \partial T}\right)_{n_j} = -RLn\gamma_i - RT\left(\frac{\partial Ln\gamma_i}{\partial T}\right)_{ni,nj}$$

Furthermore:

$$h_i^E = -T^2 \left[\frac{\partial^2\left(\frac{G^E}{T}\right)}{\partial T\, \partial n_i}\right] = -RT^2 \left[\frac{\partial}{\partial T}(Ln\gamma_i)\right]_{n_i,n_j}$$

$$v_i^E = RT\left(\frac{\partial Ln\ \gamma_i}{\partial P}\right)_{T,n_i,n_j}$$

Additionally, the overall excess enthalpy H^E is more useful than the h_i^E.

$$H^E = -RT^2 \left[\frac{\partial^2\left(G^E / RT\right)}{\partial T}\right]$$

3.1.5. Practical expressions of G^E and γ_i

Renon's expressions are highly flexible, and thus can be used to express both liquid–vapor equilibria and liquid–liquid equilibria, but it is trickier for liquid–liquid–vapor equilibria.

$$G^E = RT\Sigma x_i \left(\frac{\sum_j x_j G_{ji}\tau_{ji}}{\sum_k x_k G_{ki}}\right)$$

$$Ln\gamma_i = \frac{\sum_j x_j G_{ji}\tau_{ji}}{\sum_k G_{ki}x_k} + \sum_j \frac{x_j G_{ij}}{\sum_k G_{ki}x_k}\left(\tau_{ij} - \frac{\sum_k \tau_{kj}G_{kj}x_k}{\sum_k G_{kj}x_k}\right)$$

with:

$$C_{ij} \neq C_{ji} \text{ and } C_{ii} = 0 \quad \alpha_{ij} = \alpha_{ji}$$

$$\tau_{ij} = C_{ij} / RT \text{ and } G_{ij} = \exp(-\alpha_{ij}\tau_{ij})$$

The parameter α is dimensionless, and its value is often situated between 0.2 and 0.5. The C_{ij} values lie between -10^6 J.kmol^{-1} and 20×10^6 J.kmol^{-1}. Examples may be found in Renon [REN 71]. Remember that:

1 calorie = 4.180 joules

As regards the excess enthalpy h^E, for the parameters α_{ij} and C_{ij}, Renon [REN 71] put forward laws of linear variation as a function of the temperature. Examples can be found in his book.

NOTE.– Paris [PAR 59] offers a graphical solution to the problem of extractive distillation – i.e. involving liquid–liquid–vapor equilibria.

Very generally, distillation operations only involve one liquid phase, and in that case, the expressions proffered by Wilson [WIL 64] are often sufficient, and simpler than those given by Renon.

$$G^E = -RT\sum_i x_i Ln\left(\sum_j x_j \Lambda_{ij}\right)$$

$$Ln\gamma_i = -Ln\left(\sum_j x_j \Lambda_{ij}\right) + 1 - \sum_k \left(\frac{x_k \Lambda_{ki}}{\sum_j x_j \Lambda_{kj}}\right)$$

Most of the time, the parameters Λ_{ij} are not too far from 1, and we have:

$$\Lambda_{ij} \neq \Lambda_{ji} \text{ and } \Lambda_{ii} = 1$$

3.1.6. *Liquid–liquid equilibria*

Whilst it is totally impossible to describe liquid–liquid equilibria by the ideal solutions model, it becomes possible with the introduction of the γ_i

values. Consider two immiscible solutions A and B, at equilibrium with one another. For each component, we can write:

$$f_{Ai} = f_{Bi} \quad so \quad \gamma_{Ai} x_{Ai} \pi_i = \gamma_{Bi} x_{Bi} \pi_i$$

$\pi_i(T)$: vapor pressure of the component i at T: Pa

Thus, finally, between the two solutions A and B and for each component:

$$a_{Ai} = a_{Bi} \quad where \quad a_i = \gamma_i x_i$$

NOTE.– The calculation indicated in section 3.1.6 requires the use of an equation of state for each component, but those equations of state may be different in nature providing the values found for the chemical potentials in the pure state μ_i^+ are exact.

3.1.7. *The UNIFAC method [FRE 77]*

The UNIFAC method is a predictive method which is helpful in guiding a search in the right direction, but it is wise to confirm the results thus given with measurements when we wish to manufacture an industrial unit. This method makes use of simple concepts.

According to UNIFAC, we should predict the activity coefficients of each component of a solution on the basis of the characteristic parameters of the functional groups present in the various types of molecules in the solution.

The logarithm of the activity coefficient of component i is the sum of two terms:

– a term expressing the differences in shape and size of molecules;

– a term linked to the energy interactions.

$$Ln\gamma_i = Ln\gamma_i^S + Ln\gamma_i^E$$

In general, the first term is much smaller than the energy-related term.

3.1.8. *Concrete meaning of the activity coefficient*

Supposing the fugacity coefficient on the side of the gas is equal to 1, the liquid–vapor equilibrium is expressed by:

$$y_i P = \gamma_i \pi_i x_i$$

The apparent vapor pressure of the component i in the solution is:

$$\pi_i^{sol} = \gamma_i \pi_i^{pure}$$

Three situations are then possible:

1) $\gamma_i > 1$: the set of molecules j (where $j \neq i$), in relation to the molecules i, exert a stronger force of repulsion than the other molecules i. The molecules j tend to drive the molecules i out into the gaseous phase.

2) $\gamma_i < 1$: the set of molecules j, for each molecule i, has greater attraction than experienced by the other molecules i. They attract the molecules i from the gaseous phase toward the liquid phase and prevent them from escaping into the gaseous phase.

3) $\gamma_i = 1$: all of the molecules j behave like molecules i. This is Raoult's law for ideal solutions.

3.1.9. *Are the activity coefficients and fugacity coefficients compatible?*

We shall begin to answer this question by specifying the fugacity of a pure liquid at its boiling pressure. When the pure liquid is at equilibrium with its vapor in the pure state at saturating pressure π_i (T), we can write:

$$f^L = f^G$$

The fugacity coefficient is common to the two fluids (see section 2.3.7) and is linked to the residual Gibbs energy G^R.

$$G^R = RTLn\varphi_i^+ = v_G \pi_i - RT - RTLn\frac{\pi v_G}{RT} - \int_\infty^{v_G} \left(P - \frac{RT}{V}\right) dV$$

v_G is the molar volume of the vapor at pressure π_i (T). Thus:

$$f_i^L = \pi_i (T) \varphi_i^+ (T, \pi_i) = f_i^G$$

Hence, the fugacities are not identical to the vapor pressure π_i.

Now consider a mixture, and suppose we have an equation of state with rules of mixing.

$$f_i^G = y_i P \varphi_i^G = P \varphi_i^L = \varphi_i^L; \ \varphi_i^L = \varphi_i^L(T, P, x_i, x_j)$$

In order to find the compatibility we seek, we write:

$$f_i^L = x_i \gamma_i f_i^{Lx} :$$

$- f_i^L$: fugacity of the species i in the liquid mixture at pressure P.

$- f_i^{Lx}$: fugacity of the component i in the pure state and at its vapor pressure π_i.

More specifically, after simplification by x_i, we obtain:

$$P \varphi_i^L = \gamma_i \varphi_i^+ \pi_i$$

If this identity is satisfied, coherence is established between the calculations of equilibrium using the γ_i values and using an equation of state with rules of mixing.

3.2. Equilibrium between crystal and solution

3.2.1. *Solubility curve*

To transition from the crystal to a supercooled liquid at the same pressure and the same temperature, we can use a three-step process, examining the evolution of the enthalpy h and the entropy s.

– heating of the crystals from T to T_f (we suppose that $T_f > T$);

– melting of the crystals at T_f;

– supercooling of the liquid from T_f to T.

First step:

$$\Delta h = C_S \left(T_f - T \right) \quad \Delta s = C_S \, Ln \frac{T_f}{T}$$

Second step:

$$\Delta h = \Delta h_f \quad \Delta s = \frac{\Delta h_f}{T_f}$$

Third step (for the entropy, the heat Δh_f shifts from T_f to T):

$$\Delta h = C_L \left(T - T_f \right) \quad \Delta s = -C_L \, Ln \frac{T_f}{T}$$

The variation of the molar chemical potential along this line is (where $\mu = h - Ts$):

$$\mu_{LO} - \mu_S = \left(C_S - C_L \right) \left(T_f - T \right) - T \left(C_S - C_L \right) Ln \frac{T_f}{T} + \Delta h_f - \Delta h_f \frac{T}{T_f} \qquad [3.1]$$

Once dissolved, the solute has the chemical potential:

$$\mu_L = \mu_{L0} + RTLn\gamma x$$

At equilibrium between the crystal and the solution, we must have:

$$\mu_S = \mu_L \quad \text{so:} \quad \mu_S = \mu_{L0} + RTLn\gamma x \qquad [3.2]$$

Let us eliminate $(\mu_{L0} - \mu_S)$ between equations [3.1] and [3.2].

$$-RT \, Ln\gamma x = \Delta C \left(T_f - T \right) - T\Delta C \, Ln \frac{T_f}{T} + \Delta h_f \left(1 - \frac{T}{T_f} \right)$$

where:

$$\Delta C = C_S - C_L \left(\text{generally} > 0 \right)$$

C_S and C_L are molar specific heat capacities of the crystal S and the supercooled liquid L: $J.kmol^{-1}.K^{-1}$, respectively.

Finally, the equation of the solubility curve is:

$$RT\,Lnx = -\Delta h_f \left(1 - \frac{T}{T_f}\right) - \Delta C(T_f - T) + T\Delta C\,Ln\frac{T_f}{T} - RTLn\gamma$$

3.2.2. *Correspondence between melting and dissolution heats*

We can directly dissolve a crystal with the molar heat of dissolution $\Lambda = \Delta h_{diss}$ at temperature T.

However, we can also employ a four-step procedure:

– heating of the crystals from T to T_{melt}, which corresponds to the heat received (by the crystals):

$$C_S\left(T_{melt} - T\right)$$

– melting of the crystals: Δh_{melt}

– cooling of the liquid (we are then dealing with supermolten liquid):

$$C_L\left(T - T_{melt}\right)$$

– mixture of the solvent and the supermolten liquid at temperature T:

$$\Delta h_{disL}$$

As these two processes are equivalent, the heat of dissolution of the crystals is:

$$\Delta h_{diss} = \Delta h_{melt} + \left(C_S - C_L\right)\left(T_f - T\right) + \Delta h_{disL}$$

The heat of mixing of two liquids is written as follows, if the index i characterizes the supercooled crystal:

$$\Delta h_{dis} = h_i^E \text{ where } -\frac{h_i^E}{T^2} = \frac{\partial(g_i^E/T)}{\partial T} = \frac{\partial[(RLn\gamma_i)/T]}{\partial T}$$

The expression of Δh_{diss} is simplified if:

– the melting point T_{melt} is not too far from T;

the heat Δh_{disL} of dissolution of the supercooled liquid in the solvent is low in comparison to Δh_{melt}.

We then write:

$$\Delta h_{dissS} \# \Delta h_{melt}$$

The practical advantage to this approximation lies in the fact that the melting heats are much better known than the dissolution heats.

3.2.3. *Use of the solubility product [MEI 79]*

Consider a salt crystallizing in the form $A_{v_A}B_{v_B}nH_2O$. Its activity a is such that:

$$a^v = m_A^{v_A} \, m_B^{v_B} \gamma_{AB}^{(v_A+v_B)} a_e^n = \Pi \text{ where } v = v_A + v_B + n$$

Now consider a double salt, crystallizing in the form:

$$\alpha\left(A_{v_A} B_{v_B}\right)\beta\left(C_{v_C} D_{v_D}\right)nH_2$$

Its activity a, in solution, is such that:

$$a^v = m_A^{\alpha v_A} \, m_B^{\alpha v_B} m_C^{\beta v_C} m_D^{\beta v_D} \gamma_{AB}^{\alpha(v_A+v_B)}\gamma_{CD}^{\beta(v_C+v_D)} a_e^n = \Pi$$

$$v = \alpha\left(v_A + v_B\right) + \beta\left(v_C + v_D\right) + n$$

When the product Π of a salt, be it simple or multiple, reaches a limiting value K which is the *solubility product*, the salt in question precipitates. The solubility product is the same in all saturated solutions, regardless of the type and of the concentration of any other ions present.

If some of the water is vaporized, it is often necessary to find the vaporized mass when the solution reaches saturation (at a temperature given in advance – say, 25°C), and we also need to identify the solid phase which appears.

We then proceed by successive tests, with increasing amounts of water vaporized, each time calculating the activity of the different salts possible and comparing those activities to the corresponding solubility products.

As soon as a solubility product is reached, we see the precipitation of the salt in question.

NOTE.– Oversaturation is defined as an excess of chemical potential. More specifically, we write (see section 4.5.1):

$$\frac{\Delta\mu}{RT} = \frac{\mu - \mu^*}{RT} = v\left(Lna - Lna^*\right) = v\,Ln\frac{a}{a^*} = v\,Ln\,S = Ln\frac{\Pi}{K}$$

The solution is oversaturated if $S > 1$ or, which is tantamount to the same thing:

$$\Pi > K$$

Experience shows us that a high degree of oversaturation S is easily obtained for a solute whose solubility is low (small value of K). Highly-soluble species, on the other hand, are such that, in practical terms:

$$1.001 < S < 1.1$$

3.2.4. *Boiling delay and crystallization delay*

At constant pressure, we need to determine the gap between:

– the boiling point of a solution and the boiling point of the pure solvent. This difference is the boiling delay τ_E;

– the temperature of crystallization of the pure solvent and the temperature of crystallization of the solvent from a solution. This difference is the crystallization delay τ_C.

We shall agree that the enthalpies of vaporization Δh_v and of crystallization Δh_c do not vary significantly whether we are dealing with the pure solvent or with the solution.

Let us write the Helmholtz relation for the pure solvent and for the solution (section 1.8.4):

$$d\left(\frac{\mu_{\text{solution}}}{T}\right) = h_{\text{solution}}\, d\left(\frac{1}{T}\right) \qquad [3.3]$$

$$d\left(\frac{\mu_{\text{pure}}}{T}\right) = h_{\text{pure}}\, d\left(\frac{1}{T}\right) \qquad [3.4]$$

However, in view of the definition of the chemical potential:

$$\mu_{\text{solution}} = \mu_{\text{pure}} + RT\,Lna_e \qquad [3.5]$$

Let us subtract equation [3.4] from equation [3.3], in light of equation [3.5]:

$$d(R\,Ln\,a_e) = \left(h_{\text{pure}} - h_{\text{solution}}\right) d\left(\tfrac{1}{T}\right) = \Lambda d\left(\tfrac{1}{T}\right) \qquad [3.6]$$

Λ: molar heat of change of state (vaporizations $s > 0$; crystallizations $s < 0$).

We can integrate equation [3.6] between the pure solvent and the solution. For the pure solvent, $a_e = 1$.

$$R\,Ln\,a_e = \Lambda\left(\frac{1}{T_{\text{solut}}} - \frac{1}{T_{\text{pur}}}\right)$$

The temperature gap sought (state change time) is written:

$$\tau = T_{\text{solution}} - T_{\text{pure}}$$

Hence:

$$\tau = -\frac{RT_{\text{pure}}^2\,Ln\,a_e}{\Lambda}$$

Indeed, without major risk of error, we could take:

$$T^2 \# T_{\text{pure}}^2 \# T_{\text{pure}}\, T_{\text{sol}}$$

EXAMPLE 3.1.–

Consider a solution of salt in water at the mass fraction of 25% of NaCl. The ionic molar fraction is:

$$\frac{2\times 25 / 58.5}{\dfrac{2\times 25}{58.5} + \dfrac{75}{18}} = \frac{0.855}{0.855 + 4.17} = 0.17$$

By examining the tables, we find the activity of water and the heats of state change at 70°C and 0°C (see [ASE 99]):

	70°C (crystal)	0°C (vapor.)
$a_e \rightarrow a_w$	0.8306	0.8018
$r(J.kg^{-1})$	2.33×10^6	-0.24×10^6
$\Lambda(J.kmol^{-1})$	136.3×10^6	-14.04×10^6

Hence, the crystallization delay is:

$$\tau_{cry} = +\frac{8314\times 273.15^2 \; \text{Ln} \; 0.8018}{-14.04\times 10^6} = -9.75°C$$

and the vaporization delay:

$$\tau_{vap} = +\frac{8314\times 343.15^2 \text{Ln} 0.8306}{136.3\times 10^6} = 1.33°C$$

We can see that the crystallization delay is significantly greater (in absolute value) than the boiling delay which, itself, is nearly always less than 5°C. It is for this reason that the roads are salted in winter.

3.2.5. Expression of the activity coefficients (solid–liquid equilibria)

We have just seen that the solution of a crystal in a solvent can be considered to be a mixture of two liquids: the solvent and a supercooled liquid. Thus, as we did for mixtures of two liquids, we can use either Renon's or Wilson's expressions to express les activity coefficients.

To determine the parameters C_{ij} [REN 71] or Λ_{ij} [WIL 64], we use the vapor pressure of the solvent above the solution (the vapor pressure of super-molten crystals is negligible).

Note that Manzo *et al.* [MAN 90] used the following expressions for the activity coefficients in a binary mixture.

$$RT Ln\gamma_1 = v_1 \phi_2^2 (\delta_1 - \delta_2)^2 \quad \phi_1 = \frac{x_1 v_1}{x_1 v_1 + x_2 v_2}$$

$$RT Ln\gamma_2 = v_2 \phi_1^2 (\delta_1 - \delta_2)^2 \quad \phi_2 = \frac{x_2 v_2}{x_1 v_1 + x_2 v_2}$$

In the common literature, δ_i^2 is expressed in cal.cm^{-3}.

v_1 and v_2 are the molar volumes (cm^3.mol^{-1}). δ_1 and δ_2 are the solubility parameters.

In Beerbower *et al.* [BEE 84] is a list of solubility parameters for 60 chemical substances.

When, in solution, the crystal yields a dissociable electrolytic chemical species, the above expression must be supplemented by a term drawn from Debye and Hückel and a Bornian term [CRU 78]. This is what we shall now see.

3.2.6. *Chemical potential of an electrolyte or of an ion*

This chemical potential is defined by:

$$\mu = \mu_o + RT Lna = \mu_o + RT Ln\, \gamma x$$

For fluids, and especially in the petroleum industry, the reference state is the limiting state at evanescent pressure. In that state, all fluids are in the gaseous state and their behavior is that of a perfect gas.

Arrhenius, in 1887, supposed that ions in solution behave like independent molecules and, by analogy with perfect gases, a reference state for electrolytes and ions has been chosen. This is the limiting state, for which the concentration is evanescent.

In other words, for this state, the activity coefficient is equal to 1, just like the fugacity coefficient of a perfect gas and we set:

$$\mu = \mu_o + RT\,Ln\left(\frac{\gamma c}{c_o}\right)$$

The concentration c is measured in moles of solute per liter (or in kilomoles per m^3), and we take $c_o = 1$ mol.L^{-1}.

Certain chemists continue to use the *molality m, measured in moles per kilogram of water*, but the numbers expressing m and c remain the same, as long as we have:

mass of 1 liter of solution = 1 kilogram

3.2.7. *Equivalence, molalities and molar fractions*

In a liter of water, there are 55.51 moles of water, and the molar fraction of solute (of which m moles are present) is:

$$x = \frac{m}{55.51 + m_T} = \frac{m}{55.51}\left(\frac{1}{1 + 0.018 m_T}\right)$$

m_T is the sum of the molalities of all the dissolved species.

Hence:

$$RT\,Lnx = RT\,Ln\,m - RT\,Ln\,55.51 + \Gamma_G$$

with:

$$\Gamma_G - RT Ln(1 + 0.018\ m_T)$$

The chemical potential μ_j is:

$$\mu_j = \mu_{jox} + RT\,Ln\,x_j\,\gamma_j \quad a_j = \overline{\gamma}_j m_j$$

$$\mu_j = \mu_{jox} + RT\,Ln\,m_j + \Gamma_G - RT\,Ln\,55.51 + RTLn\,\gamma_j$$

Let us set:

$$\mu_{jom} = \mu_{jox} - RT\,Ln\,55.51$$

Hence, the chemical potential based on the molalities is:

$$\mu_j = \mu_{jom} + \Gamma_G + RT\,Ln\,m_j + RT\,Ln\,\gamma_j$$

3.2.8. Average activity

Suppose that a salt dissociates in accordance with:

$$M_m A_n \rightarrow mM^+ + nA^-$$

Let us posit:

$$\nu = m + n \text{ and } a_\pm = a^{1/\nu} \text{ and } \gamma_\pm = \left[\gamma_{M^+}^m \gamma_{A^-}^n\right]^{1/\nu}$$

However, if x is the molar fraction of the salt in the solution:

$$x_{M^+} = mx \quad \text{and} \quad x_{A^-} = nx$$

and also set:

$$x_\pm = \left(x_{M^+}^m . x_{A^-}^n\right)^{1/\nu} = x\left(m^m . n^n\right)^{1/\nu}$$

Hence:

$$a_\pm = x_\pm \gamma_\pm \text{ and } a_{M^+}^m . a_{A^-}^n = a_\pm^\nu$$

3.3. Electric field of Debye and Hückel [DEB 23]

3.3.1. Correspondence between the international system and the electrostatic CGS system

In a vacuum, the force F exerted between two unitary electrical charges e, separated by the distance r, is such that:

$$Fr^2 = Ke^2$$

In the electrostatic CGS system (see Appendix N):

$$F \times r^2 = e_{es}^2 = \left(4.8029 \times 10^{-10}\right)^2 = 2.3067 \times 10^{-19} \text{ dyne.cm}^2.$$

International system:

$$F \times r^2 = \frac{e^2}{4\pi\varepsilon_0} = \frac{(1.60206 \times 10^{-19})^2}{4\pi \times 8.854 \times 10^{-12}} = 2.3067 \times 10^{-28} \text{ N. m}^2$$

We can verify that:

$$\frac{\text{Newton} \times \text{m}^2}{\text{dyne} \times \text{cm}^2} = \frac{10^{-19}}{10^{-28}} = 10^9$$

NOTE.– 1) From the dimensional point of view, it is necessary to write:

$$\left[e^2\right] = \frac{F \times L^2}{\text{molecule}}$$

This enables us to quickly find the dimension of the Debye length and those of the excess Helmholtz electrical energy F_e^E.

2) The variation of the dielectric constant of water as a function of the temperature can be found in Archer *et al.* [ARC 91]. Note that, at 1 bar abs. and at 25°C, that constant (which is none other than the relative permittivity of water) is:

$$\varepsilon_{1;25} = 78.38$$

3.3.2. *Debye length*

The equation for the electrical field, according to the authors, is:

$$\Delta E = \kappa^2 E$$

From this, Debye and Hückel deduce the *Debye length* $1/\kappa$, defined by:

$$\kappa^2 = \frac{4\pi e^2}{\varepsilon kT}\Sigma c_i z_i^2, \text{ and in international units } \kappa^2 = \frac{e^2}{\varepsilon\varepsilon_0 kT}\Sigma c_i z_i^2 :$$

– ε : relative permittivity of water.

– c_i : concentration: molecules.cm^{-3}.

– e: charge on the electron.

One mole per liter (or 1 kmol.m^{-3}) is equivalent to $(N_A/1000)$ molecules per cm^3.

Let us set:

$$B_\gamma^2 = \frac{8\pi e^2}{\varepsilon kT} \times (N_A / 1000) \text{ (CGS)}$$

and:

$$B_\gamma^2 = \frac{8\pi e^2 N_A}{4\pi\varepsilon\varepsilon_0 kT} = \frac{2 e^2 N_A}{\varepsilon\varepsilon_0 kT} \text{ (S.I.)}$$

The Debye length (or rather, its inverse):

$$\kappa = \left(B_\gamma^2 I\right)^{1/2} = B_\gamma I^{1/2}$$

EXAMPLE 3.2.–

Let us calculate the coefficient B_γ in both systems of units (at 25°C).

CGS system:

$$B_\gamma^2 = \frac{8\pi(4.8029\times10^{-10})^2 \times 6.02252\times10^{20}}{78.38\times1.38053\times10^{-16}\times298.15} = 1.0822\times10^{15}\, \text{cm}^{-2}(\text{mol.L}^{-1})^{-1}$$

International system (see Appendix 8):

$$B_\gamma^2 = \frac{2\times(1.60206\times10^{-19})^2 \times 6.02252\times10^{26}}{78.38\times8.85434\times10^{-12}\times1.38054\times10^{-23}\times298.15}$$

$$B_\gamma^2 = 1.0822 \times 10^{19} \ m^{-2} (kmol.m^{-3})^{-1}$$

Hence, at 25°C:

$$B_y = 0.3289 \times 10^8 \ cm^{-1} (mol.L^{-1})^{-1/2} = 0.3289 \times 10^{10} \ m^{-1} (kmol.m^{-3})^{-1/2}$$

3.3.3. *Ionic radius and associated coefficients*

The values of some ionic radii are to be found in Table 3 of Helgeson *et al.* [HEL 81].

To calculate the Helmholtz energy, and therefore the ionic activity coefficients, Debye and Hückel bring in a coefficient χ_j defined thus:

$$\chi_j = \frac{3}{\alpha_j^3} \left[\frac{3}{2} + Ln(1+\alpha_j) - 2(1+\alpha_j) + \frac{1}{2}(1+\alpha_j)^2 \right]$$

α_j is given by:

$$\alpha_j = \kappa r_j \ :$$

κ: inverse of the Debye length: m^{-1}(S.I.) or cm^{-1} (CGS)

r_j: ionic radius in m (S.I.) or cm (CGS).

Eyring, in his book [EYR 64], replaces the coefficient χ_j with the fraction $1/(1+\alpha_j)$, with no justification. We shall not employ this modification, which leads to an error of around 10%.

To calculate the osmotic coefficient of water – or its activity coefficient, which amounts to the same thing – Debye and Hückel employ a coefficient σ_j for each ion present in the solution.

$$\sigma_j = \frac{3}{\alpha_j^3} \left[(1+\alpha_j) - \frac{1}{1+\alpha_j} - 2Ln(1+\alpha_j) \right]$$

3.3.4. *Helmholtz energy of the solution [DEB 23]*

According to Debye and Hückel's calculations, the electrical Helmholtz energy for the solution is:

$$F_e = -\frac{kT}{4\pi\Sigma c_i z_i^2}\Sigma_i n_i z_i^2 \int \frac{\kappa^2 d\kappa}{1+\kappa a_i} \quad \text{(CGS)}$$

n_i: number of ions of the species i

a_i: radius of the ion i: cm

After integration, we obtain (CGS):

$$F_e = -\frac{\Sigma n_i z_i^2 e^2 \kappa\chi_i}{3\varepsilon} \quad \text{(erg)}$$

In the international system:

$$F_e = -\frac{\Sigma n_i z_i^2 e^2 \kappa\chi_i}{12\pi\varepsilon\varepsilon_o} \quad \text{(J)}$$

3.3.5. *Partial derivatives of F_e*

The authors calculate, for neutral molecules (solvent or otherwise):

$$\frac{\partial(\kappa\chi_i)}{\partial n_j} = \frac{d(\kappa\chi_i)}{d\kappa} \times \frac{d\kappa}{dn_j}$$

Here, j is the typical index of neutral molecules:

$$\frac{\partial\kappa}{\partial n_j} = -\frac{\kappa\overline{v}_j}{2V} \quad \text{where} \quad V = \sum_i n_i\overline{v}_i + \sum_j n_j\overline{v}_j$$

$\overline{v}_j$: partial molar volume of the species j: cm^3.mol^{-1}

$$\frac{d(\kappa\chi_i)}{d\kappa} = \sigma_i = \frac{3}{\alpha_i^3}\left[1+\alpha_i - \frac{1}{1+\alpha_i} - 2\text{Ln}(1+\alpha_i)\right]$$

Finally, if F_e is the electrical Helmholtz energy:

$$\frac{\partial F_e}{\partial n_j} = \overline{v}_j \sum_i \frac{\kappa c_i z_i^2 e^2 \sigma_i}{6\epsilon} = \frac{\overline{v}_j}{V} \sum_i \frac{\kappa n_i z_i^2 e^2 \sigma_i}{6\epsilon} \quad \text{(CGS)}$$

For ions, we need to find the sum of the above derivative:

$$\left(\frac{\partial F_e}{\partial n_i}\right)_1 = \frac{\kappa e^2}{6\epsilon} \frac{\overline{v}_i}{2} \sum_k c_k z_k^2 \, \sigma_k$$

and:

$$\left(\frac{\partial F_e}{\partial n_i}\right)_2 = -\frac{\kappa e^2}{3\epsilon} z_i^2 \chi_i$$

Thus, in total:

$$\frac{\partial F_e}{\partial n_i} = -\frac{\kappa e^2}{3\epsilon}\left[z_i^2 \chi_i - \overline{v}_i \sum_k \sigma_k \frac{c_k z_k^2}{2} \right]$$

It is seemingly more realistic to replace χ_i and σ_k with mean values obtained on the basis of a mean value of the α_i coefficients.

As the ionic strength is:

$$I = \frac{1}{2} \sum_k c_k z_k^2$$

the derivative becomes:

$$\frac{\partial F_e}{\partial n_i} = -\frac{\kappa e^2}{3\epsilon}\left(\chi z_i^2 - \overline{v}_i \sigma I\right) \quad \text{(CGS)}$$

Naturally:

$$\frac{\partial F_e}{\partial n_j} = kTLn\gamma_j \quad \text{et} \quad \frac{\partial F_e}{\partial n_i} = kTLn\gamma_i$$

NOTE.– When the concentration of an electrolyte tends toward zero, Debye and Hückel's electrical field disappears, and the corresponding activity coefficient tends toward 1 because $\partial F_e/\partial n_i$ tends toward zero. This is not a convention: it is a physical reality.

3.4. Two approximate calculation methods (electrolytes) [MEI 80] and [HEL 81]

3.4.1. *Some definitions to begin with*

A solution may contain charged atoms, but also complexes of multiple atoms, whether or not those complexes carry a charge.

The total ionic strength of the solution is:

$$\overline{I} = \sum_j \psi_j m_j + \sum_q \psi_q m_q$$

$$\psi_j = \frac{z_j^2}{2} \quad \psi_q = \frac{z_q^2}{2} \quad \text{(equation 84 in the author's text)}$$

z_j and z_q are the electrical valences, respectively, of the atoms and complexes, charged or otherwise (in the latter case, $z_q = 0$).

The atomic ionic strength pertains only to charged atoms:

$$I = \sum_j \psi_j m_j$$

The total molality of the solution is defined by the sum of the molalities of the ions present in the solution (see section 3.2.6):

$$m_T = \sum_j m_j$$

3.4.2. *Ionic strength and valences*

Suppose an electrolyte dissociates into ν_1 ions of valence z_1 and ν_2 ions of valence z_2. The index 1 might, for example, characterize a cation, and the index 2 an anion.

The law of electrical neutrality means that, with z_1 and z_2 being counted positively, we can write:

$$v_1 z_1 = v_2 z_2 \qquad\qquad [3.6b]$$

Furthermore, the ionic strength I corresponding to a dissociated molecule is, by definition, given by:

$$2I = v_1 z_1^2 + v_2 z_2^2$$

which, in light of equation [3.6b], can be written:

$$2I = v_1 z_1^2 + v_2 z_2^2 = \frac{v_1^2 z_1^2}{v_1} + \frac{v_2^2 z_2^2}{v_2} = v_1 z_1 v_2 z_2 \left(\frac{1}{v_1} + \frac{1}{v_2} \right) = z_1 z_2 \left(v_1 + v_2 \right) = z_1 z_2 v_s$$

3.4.3. *Activity of water (single electrolyte)*

As we know that one liter of water contains 55.51 moles, the Gibbs–Duhem relation (see section 1.8.1 and 1.8.2) is written:

$$55.51 d \, Ln \, a_e + \left(v_1 + v_2 \right) m_s \, d \, Ln \left(m_s \gamma_\pm \right) = 0 \qquad\qquad [3.7]$$

Indeed, the concentration c_s of the solute (moles per liter), which is expressed by the same number as the molality, corresponds to $(v_1 + v_2) \, m_s$ ions after dissociation.

The ionic strength of the dissolved and dissociated electrolyte is:

$$I_s = \frac{1}{2} (v_1 z_1^2 + v_2 z_2^2) m_s = \frac{1}{2} z_1 z_2 v_s m_s$$

Let us differentiate:

$$0.5 v_s dm_s = \frac{dI}{z_1 z_2}$$

We set:

$$\frac{Ln \gamma_\pm}{z_1 z_2} = Ln \, \Gamma_s$$

Γ_s is the *reduced* activity coefficient.

Equation [3.7] can be written:

$$27.75\,d\,Ln\,a_e + 0.5\,v_s\,m_s\,d\,Ln\,m_s + 0.5\,v_s\,m_s\,d\,Ln\,\gamma_\pm = 0$$

This means that:

$$27.75\,d\,Ln\,a_e + \frac{dI_s}{z_1 z_2} + I_s\,d\,Ln\,\Gamma_s = 0$$

Now, if we integrate:

$$27.75\,Ln\,a_e = \frac{-I_s}{z_1 z_2} - \int_0^{\Gamma s} I_s\,d\,Ln\,\Gamma_s$$

This calculation can be performed if we know the function Γ_s (I_s). Meissner [MEI 80] put forward such a relation (see section 3.4.4).

3.4.4. *Meissner's equation [MEI 80]*

For a given electrolyte, the author posits that:

$$Ln\,\Gamma^* = -\frac{1.1745\sqrt{I_s}}{1 + c\sqrt{I_s}}$$

where:

$$c = 1 + 0.055\,q\,\exp\left(-0.023\,I_s^3\right)$$

Thus, for a single dissolved electrolyte (*pure* solution), we obtain:

$$\Gamma_s^0 = \Gamma^*\left[1 + B(1 + 0,1I)^q - B\right]$$

where:

$$B = 0.75 - 0.065\,q$$

The parameter q is characteristic of the electrolyte. Kusik and Meissner [KUS 78] give the value of parameter q for 121 salts commonly used in industry.

EXAMPLE 3.3.–

Activity coefficient of $MgCl_2$ for an ionic strength $I = 0.07$.

For this salt:

$q = 2.90$

$$c = 1 + 0.055 \times 2.9 \exp\left(-0.023 \times 0.07^3\right) = 1.159498$$

$$\text{Ln } \Gamma^* = -\frac{1.1745 \times 0.07^{1/2}}{1 + 1.159498 \times 0.07^{1/2}} = -0.238083$$

$\Gamma^* = 0.788137$ and $B = 0.75 - 0.065 \times 2.9 = 0.5615$

Here, we are dealing with a *pure* solution:

$$\Gamma_s^\circ = 0.788137\left[1 + 0.5615\left(1 + 0.1 \times 0.07\right)^{2.9} - 0.5615\right]$$

$$\Gamma_s^\circ = 0.797180$$

$$\gamma_\pm = \left(\Gamma_s^\circ\right)^2 = 0.635496$$

Note that:

$$\log_{10} \gamma_\pm = -0.196886$$

3.4.5. *Meissner curves*

In Meissner's paper [MEI 80], we find a lattice of curves (see Figure 2) representing the variations of Γ_s as a function of the ionic strength and the parameter q. In addition, a lattice of transverse curves represents constant values of the activity of water $a_e \rightarrow a_w$ *for a 1–1 electrolyte*.

Remember that the activity coefficient of the solute is:

$$\text{Ln } \gamma_{\pm} = z_1 z_2 \text{ Ln } \Gamma_s$$

The osmotic coefficient and the osmotic pressure are:

$$\phi = -\frac{\text{Ln } a_e}{V_e \, v_s \, m_s} \qquad \Pi = -\frac{RT \, \text{Ln } a_e}{V_e} = RT \, \phi v_s m_s$$

For an electrolyte containing ions whose valences are greater than 1, we operate as follows.

We write the equation of the activity of water for a fictitious 1–1 electrolyte wherein the value of q is the same as that of the electrolyte $z_i - z_\ell$. We also write that equation for the electrolyte $z_i - z_\ell$. The integrals of these two equations are identical because they are calculated over the same interval of ionic strength.

By subtracting the two equations from one another, term by term, the integrals are eliminated, and we obtain:

$$27.75 \text{ Ln}(a_e)_{zi:zl} = I \times \left(1 - \frac{1}{z_i z_l}\right) + 27.75 \text{ Ln}(a_e)_{1:1}$$

Meissner's lattice of curves can be used to determine the reduced activity coefficient if we know the ionic strength, and the activity a_e -> a_w of the solvent.

3.4.6. *Reduced activity coefficient in a mixture*

Solid salts whose ions have multiple charges (valence greater than 1) are almost always hydrated if they are in contact with their saturated solution.

The interaction between 2 ions *whose charge is of the same sign is negligible.*

The excess Helmholtz energy corresponding to the interaction between the cation i_o and the anions of common index ℓ is proportional to:

$$F_{i_0} = \frac{1}{2}\sum_l Y_l \text{Ln}\Gamma_{i_0 l}^o \quad \left(\text{with } \Gamma_l = \frac{\gamma_l}{z_i z_l}\right)$$

where:

$$Y_l = \frac{1}{2}m_l z_l^2 / I_L \text{ and } I_L = \frac{1}{2}\sum_l m_l z_l^2$$

I_L is the total ionic strength of the anions.

The $\Gamma_{i_o l}^o$ are calculated for a pure solution (containing only the salt $i_o \ell$) *on the basis of the total ionic strength* I_T of the solution:

$$I_T = \frac{1}{2}\left(\sum_i m_i z_i^2 + \sum_l m_l z_l^2\right)$$

Symmetrically, for the anion ℓ:

$$F_{lo} = \frac{1}{2}\sum_i X_i Ln\Gamma_{ilo}^o \text{ with } X_i = \frac{1}{2}m_i z_i^2 / I_l \text{ and } I_l = \frac{1}{2}\sum_i m_i z_i^2$$

I_l is the total ionic strength of the cations.

The rule dictates that the reduced activity coefficient of the salt $i_o \ell_o$ in the mixture be such that:

$$Ln\Gamma_{i_o l_o} = F_{i_o}F_{l_o} = \frac{1}{2}\sum_i X_i Ln\Gamma_{ilo}^o + \frac{1}{2}\sum_l Y_l Ln\Gamma_{iol}^o$$

This relation applies whether or not the solution is saturated. It also applies whatever the form of the solid phase at equilibrium with the solution. In their 1972 publication [MEI 72a], Meissner and Kusik draw the consequences of that law of mixing.

In their 1973 paper [MEI 73b], Meissner and Peppas more specifically examine the case of acids, and indicate that polyacids often behave like monoacids because the first dissociation is significant on its own.

3.4.7. *Activity of the solvent in a mixed solution*

Unlike with a "pure" solution, a so-called "mixed" solution contains at least three different types of ions. Having abandoned the expression of $a_e \rightarrow a_w$ given in Meissner and Kusik [MEI 78] in equation (1), in 1980, Meissner proposed operating in the following manner:

The parameter $q_{i_o l_o}$ is given by:

$$qi_o l_o = \sum_l \frac{I_l}{I_T} qi_{ol} + \sum_i \frac{I_i}{I_T} qil_o \quad \text{where } I_T = \frac{1}{2}\sum_i m_i z_i^2 + \frac{1}{2}\sum_l m_l z_l^2$$

We can then calculate the fictitious activity $A_{i_o l_o \text{mix}}$ of the electrolyte $i_o \ell_o$ by using the parameter $q_{i_o l_o}$ in the equation:

$$27.75 \, LnA_{i_o l_o \text{mix}} = -\frac{I_T}{z_{i_o} z_{l_o}} - \int_0^{I_T} I_{i_o l_o} dLn\Gamma_{i_o l_o}$$

We can also use the curves given by Meissner [MEI 80] with $q_{i_o l_o}$ and I_T.

I_T: total ionic strength of the mixture: $mol.L^{-1}$ (or $kmol.m^{-3}$).

Now consider $x_{i\ell}$, the molar fraction of the electrolyte $i_o \ell_o$ in the set of electrolytes dissolved to form the solution.

The activity of the solvent in the solution is then:

$$a_{\text{emix}} = \prod_{i_o} \prod_{l_o} A_{i_o l_o \text{mix}}^{x_{i_o l_o}}$$

3.4.8. *Water: an ionizing and solvating solvent*

Water is an *ionizing* solvent, because of its high dielectrical relative permittivity constant which decreases the electrical interactions between the ions. This constant is around 80 (see note (2) in section 3.3.1).

However, water is also a *solvating* solvent, which means that the ions are surrounded by a coating of water molecules. More specifically, cations whose valence is greater than 1 are always solvated, because the oxygen atom (negative) in water is attracted by the cation, which carries a positive charge. Solvation of the anions is also possible, but to very varying degrees.

The result of solvation is that crystals containing metal ions whose valence is greater than 1 are always hydrated.

In general, solvation is strongly exothermic:

$$-4500 \, MJ.kmol^{-1} < \Delta H_{\text{solv}} < -400 \, MJ.kmol^{-1}$$

3.4.9. *Influence of temperature*

The above results pertain to solutions at 25°C. Meissner [MEI 80] proposed a method to determine the parameter q° at a temperature t°C.

$$q_t^\circ = q_{25°C}^\circ \left[1 - \frac{0.0027(t-25)}{z_1 z_2}\right]$$

Remember that the parameter q° relates to a single electrolyte in solution (i.e. a *pure* solution).

Depending on the sign of $q_{25°C}^\circ$, q_t° increases or decreases with t.

The author recommends this procedure for temperatures such that:

25°C < t < 120°C

3.4.10. *Criticism of Debye and Hückel's activity coefficient*

According to Debye and Hückel, the excess Helmholtz energy of electrical origin is written (at constant volume):

$$F_e = -\sum_i \frac{n_i z_i^2 e^2 \kappa \chi_i}{3\varepsilon} \quad \text{(CGS)}$$

However:

$$\kappa = \frac{2\sqrt{\pi}\, e \left(\Sigma c_i z_i^2\right)^{1/2}}{(\varepsilon kT)^{1/2}} = \frac{2\sqrt{\pi}\, e}{(\varepsilon VkT)^{1/2}} \left(\Sigma_i n_i z_i^2\right)^{1/2}$$

where:

$$c_i = \frac{n_i}{V}$$

V: volume of the solution.

Hence:

$$F_e = -\frac{2\sqrt{\pi}\, e^3 \chi_i}{3\varepsilon^{3/2} (VkT)^{1/2}} \left(\Sigma_i n_i z_i^2\right)^{3/2}$$

$$\frac{\partial F_e}{\partial n_i} = -\frac{2\sqrt{\pi}\, e^3 \chi_i}{3\varepsilon^{3/2} (VkT)^{1/2}} \times z_i^3 \times \frac{3}{2}\left(\sum_i n_i z_i^2\right)^{1/2} z_i^2$$

However:

$$\sum_i n_i z_i^2 = 2V\left(\frac{1}{2}\sum_i c_i z_i^2\right) = 2\,V\,I$$

$$\mu_{ei} = \frac{\partial F_e}{\partial n_i} = -\frac{\sqrt{2\pi}\, e^3 \chi_i I^{1/2} z_i^2}{\varepsilon^{3/2} (kT)^{1/2}} \quad \text{(CGS)}$$

or indeed:

$$\text{Ln}\,\gamma_i = -\frac{\sqrt{2\pi}\, e^3 \chi_i I^{1/2} z_i^2}{\varepsilon^{3/2} (kT)^{3/2}} \quad \text{(CGS)}$$

The number of molecules contained in a cm^3 is equal to the number of moles contain in a liter multiplied by Avogadro's number over 1000:

$$\text{Ln}\,\gamma_i = -\frac{\sqrt{2\pi}\, e^3 \chi_i (N_A /1000)^{1/2} I^{1/2} z_i^2}{\varepsilon^{3/2} kT^{3/2}} = A_\gamma \left(I^{1/2} z_i^2 \chi_i\right)$$

In the international system, we simply need to multiply the ionic strength I by Avogadro's number N_A, and we have to divide by $(4\pi\varepsilon_o)^{3/2}$ because the electronic charge is to the power of 3.

$$\text{Ln}\gamma_i = \frac{e^3 N_A^{1/2}}{2\pi(2\varepsilon\varepsilon_o kT)^{3/2}} \times \left(I^{1/2} z_i^2 \chi_i\right)$$

$$\text{Ln}\gamma_i = -A_\gamma\left(I^{1/2} z_i^2 \chi_i\right)$$

EXAMPLE 3.4.–

Calculation of A_γ

CGS system:

$$A_\gamma = \frac{\sqrt{2\pi}\left(4.8029\times10^{-10}\right)^3 \times \left(6.02252\times10^{20}\right)^{1/2}}{\left(78.38\times1.38053\times10^{-16}\times298.15\right)^{3/2}}$$

$$A_\gamma = 1.1761$$

International system:

$$A_\gamma = \frac{\left(1.60206\times10^{-19}\right)^3\times\left(6.02252\times10^{26}\right)^{1/2}}{2\pi\left(2\times78.38\times8.85434\times10^{-12}\times1.38054\times10^{-23}\times298.15\right)^{3/2}}$$

$$A_\gamma = 1.1761$$

3.4.11. Activity coefficient of the solvent

According to Debye and Hückel [DEB 23], the activity coefficient of water is obtained by:

$$kT\,Ln\,\gamma_e = -\frac{e^2\kappa}{6\varepsilon}\sum_i\sigma_i x_i z_i^2$$

Certain authors [HEL 81] adopt a mean value for σ_i so as to make the ionic strength apparent. This mean value is calculated on the basis of a mean ionic radius of the solution:

$$\overline{r} = \frac{2}{c_T}\sum_j^s c_j r_j \quad \left(c_T = \sum_j c_j\right):$$

– c_T: number of ions – moles in the unit volume of the solution.

– c_j: number of ions of type j in the unit volume of the solution.

We therefore obtain:

$$kT\,Ln\,\gamma_e = \frac{e^2\kappa\sigma\times\sum_i x_i z_i^2}{6\varepsilon}$$

Let us multiply and divide by the total concentration expressed in molecules per cm^3 and, in addition:

$$\frac{1}{2}\sum_i c_T x_i z_i^2 = \frac{1}{2}\sum_i c_i z_i^2 = I \quad \left(c_T = m^*\right)$$

Hence, for the activity coefficient γ_e of water:

$$kT \, Ln \, \gamma_e = -\frac{e^2 \kappa \sigma}{3\varepsilon} \frac{I}{c_T}$$

However, we know that:

$$\kappa = \left(\frac{2I \times 4\pi e^2}{\varepsilon kT}\right)^{1/2}$$

Hence:

$$Ln \, \gamma_e = \frac{e^3 (8\pi)^{1/2} \sigma I^{3/2}}{3(\varepsilon kT)^{3/2} \times C_T}$$

Yet:

$$\frac{I^{3/2}}{c_T} (\text{molecules. cm}^{-3})^{1/2} = \left(\frac{N_A}{1\,000}\right)^{1/2} (\text{moles. L}^{-1})^{1/2}$$

Hence:

$$Ln \gamma_e = -\frac{e^3 (8\pi N_A/1\,000)^{1/2}\sigma}{3(\varepsilon kT)^{3/2}} \times \frac{I^{3/2}}{c_T} \quad (\text{CGS systems})$$

To switch to the S.I., we need to divide by $(4\pi\varepsilon_0)^{3/2}$ and express e in coulombs. In this system, division by 1000 is no longer necessary. We obtain:

$$Ln \, \gamma_e = -\frac{e^3 (2N_a)^{1/2}\sigma}{12\pi(\varepsilon\varepsilon_0 kT)^{3/2}} \times \frac{I^{3/2}}{c_T} \quad (\text{S.I})$$

EXAMPLE 3.5.–

1) CGS:

$$e = 4.8029 \times 10^{-10} \quad N_A = 6.02252 \times 10^{23}$$

$$T = 298.15 \quad kT = 4.116 \times 10^{-14} \, erg \quad \varepsilon = 80$$

$$\frac{\left(4.8029\times10^{-10}\right)^3\left(8\pi\times6.02252\times10^{20}\right)^{1/2}}{3\times\left(80\times4.116\times10^{-14}\right)^{3/2}}=0.76$$

$$Ln\gamma_e = 0.76\frac{\sigma I^{3/2}}{c_T}$$

2) S.I.:

$$e = 1.60206\times10^{-19}\ N_A = 6.02252\times10^{26}\ kT = 4.116\times10^{-21}J$$

$$\varepsilon_0 = 8.854\times10^{-12}$$

$$\frac{\left(1.60206\times10^{-19}\right)^3\left(2\times6.02252\times10^{26}\right)^{1/2}}{12\pi\left(80\times8.854\times10^{-12}\times4.116\times10^{-21}\right)^{3/2}}=0.76$$

Indeed, we find:

$$Ln\ \gamma_e =- 0.76\frac{\sigma I^{3/2}}{c_T}$$

Here, I and c_T are expressed in $kmol.m^{-3}$.

3.4.12. *Activity coefficients of solutes according to Helgeson et al. [HEL 81]*

We refer to these authors' tables and equations by their marker (number) in their 1981 publication [HEL 87].

For the ion j, the activity coefficient of an ion is given by their equation 196:

$$\log_{10}\overline{\gamma}_j = -\frac{A_\gamma z_j^2 \overline{I}^{1/2}}{\Delta}+\Gamma_\gamma + \omega_j^{abs}\sum_k b_k y_k \overline{I}+\sum_p b_{pj}m_p$$

The bar above γ_j and I characterizes the real property (incomplete dissociation) of the property given by stoichiometry (complete dissociation).

Hereinafter, we shall not include it, because we are looking at fully-dissociated electrolytes, and can simply write:

$$\log_{10}\gamma_j = -\frac{A_\gamma z_j^2 I^{1/2}}{\Delta} + \Gamma_\gamma + \omega_j^{abs}\sum_k b_k y_k I + \sum_p b_{pj}m_p$$

$$\Delta = 1 + \mathring{a}_h B_j I^{1/2}$$

The parameter å, in angströms (10^{-10} m), measures the minimum distance to which the ions may approach the electrolyte k. In reality, $\mathring{a}_k$ is a mean value, because the electrolyte k is dissociated.

$$\mathring{a}_k = \left(2\sum_j v_{jk}r_j\right)/v_k \quad \text{where} \quad v_k = \sum_j v_{jk}$$

r_j is the radius of an ion j in solution. It is given by Table 3 of the authors.

$\mathring{a}_k$ is given by the authors' Table 2.

I is the ionic strength:

$$I = \frac{1}{2}\sum_j z_j^2 m_j$$

B_γ is given by the authors' Table 1, as is the coefficient A_γ (see section 3.4.10).

$$\Gamma_\gamma = -\log_{10}\left(1 + 0.0180153\,m^*\right)$$

The molality m_T is the total molality in terms of ions. If the solution contains only one electrolyte whose molality is m_k, we have:

$$m = v_{ik}m_k + v_{lk}m_k$$

According to the authors' equations 82 to 89:

$$y_k I = \psi_k m_k = \sum_j v_{jk}\frac{z_j^2}{2}m_k$$

and:

$$\sum_k b_k y_k I = \sum_k b_k \sum_j v_{jk} z_j^2 \frac{m_k}{2} = \sum_k \left[b_k m_k \sum_j \left(\frac{v_{jk} z_j^2}{2} \right) \right]$$

In addition, for the term ω_j involved in Born's equation (see section 3.6.2), we can write:

$$\omega_j = \omega_j^{abs} - z_j \omega_{H^+}^{abs} \quad (cal.mol^{-1})$$

However:

$$\sum_j v_{jk} z_i = 0 \quad \text{(electrical neutrality)}$$

Thus, it makes no difference whether we use the ω_j^{abs} or the ω_j in calculating the $\gamma_\pm$ of an electrolyte. Indeed:

$$\omega_k = \sum_j v_{jk} \omega_j^{abs} = \sum v_{jk} \omega_j$$

where:

$$\omega_j^{abs} = \frac{N_A e^2 z_j^2}{2r_{ej}}$$

$-$ r_{ej}: electrostatic ionic radius in solution given by the author's Table 3:

$$r_{ej} = r_{xj} + |z_j| \Gamma_z$$

- cations $\Gamma_z = 0.94$

- anions $\Gamma_z = 0$

$-$ r_{xj}: ionic radius of crystallization given by the author's Table 3.

However, for the calculation of the activity coefficient of an isolated ion, we are obliged to use ω_j^{abs}.

The coefficient b_k is given by the authors' Table 4. It is a solvation parameter and is expressed in L.cal^{-1}, so the product $\omega_j b_k$ is expressed in L.mol^{-1}. It is the inverse of the dimensions of an ionic strength.

The coefficient b_{pj} expresses the short-range actions on the ion j of ions whose charge is of the opposite sign to that of the ion j. The values of the b_{pj} coefficients are given by the authors' Table 7.

The mean activity coefficient of the electrolyte is then given by:

$$\log_{10} \gamma_{\pm} = \frac{1}{v_k} \sum_j v_{jk} \log_{10} \gamma_j$$

NOTE.– It is important to be able to calculate the individual activity coefficient of an ion, because that value *plays a part in the diffusion of more than two ions* – particularly in electrolysis and in transport of material (see section 4.5.3).

EXAMPLE 3.6.–

We shall calculate the activity coefficient of magnesium chloride for an ionic strength equal to 0.07 mole.L^{-1}, at 25°C.

$$A_{\gamma} = 0.5091 \qquad B_{\gamma} = 0.3283 \qquad \mathring{a} = 4.11$$

$$\Lambda = 1 + 4.11 \times 0.3283 \times (0.07)^{1/2} = 1.356893$$

$$I = 0.07 = \frac{1}{2}\left(v_{Cl^-} z_{Cl^-}^2 + v_{Mg} z_{Mg}^2\right) m_{MgCl_2} = \frac{1}{2}(2+4) m_{MgCl_2}$$

$$m_{MgCl_2} = 0.02333....$$

$$m_T = 3 \times 0.0233 ... = 0.07$$

$$\Gamma_{\gamma 10} = -\log_{10}(1 + 0.0180153 \times 0.07) = -0.0005468$$

$$b_{MgCl_2} = 1.3291 \times 10^{-6}$$

1) Activity coefficient of chlorine:

$$b_{MgCl_2} = 1.3291 \times 10^{-6}$$

$$b_{pj} = -0.256$$

$$1.3291.10^{-6} \times 0.02333 \times \frac{(2+4)}{2} = 93.0368 \times 10^{-9}$$

$$\log_{10} \gamma_{Cl^-} = -\frac{0.5091 \times 0.264575}{1.356893} - 0.0005468 + 1.4560 \times 10 - {}^5 \times 93.0368 \times 10^{-9}$$

$$- 0.256 \times 0.0233...$$

$$\log_{10} \gamma_{Cl^-} = -0.0922414$$

2) Activity coefficient of magnesium:

$$\omega_j = 1.5372 \times 10^5$$

$$b_{pj} = -0.256$$

$$b_k m_k \sum_j \frac{v_{jk} z_j^2}{2} = 93.0368 \times 10^{-9}$$

$$\log_{10} \gamma_{Mg^{++}} = -\frac{0.5091 \times 4 \times 0.07^{1/2}}{1.356893} - 0.000546865$$

$$+1.53 + 2 \times 10^5 \times 93.0368 \times 10^9 - 0.256 \times 0.04666...$$

$$\log_{10} \gamma_{Mg^{++}} = -0.39526$$

3) Average coefficient of $MgCl_2$:

$$\log_{10} \gamma_{MgCl_2} = -\frac{1}{3}(0.0922414 \times 2 + 0.39526)$$

$$\log_{10} \gamma_{MgCl2} = -0.193$$

We can see that this result is very similar to that obtained by Meissner's method, which was:

$$\log_{10} \gamma_{MgCl_2} = -0.196$$

NOTE.– The method employed by Bromley [BRO 73], and that of Pitzer and Mayorga [PIT 73, MAY 74], yield more widely dispersed results. For this reason, we have not used them here.

3.4.13. *Influence of temperature [HEL 81]*

As we have seen, three parameters play a part in calculating the activity coefficient:

$$\omega_j \left(and \, \omega_k = \sum_j v_{jk} \omega_j \right), b_k \text{ and } b_{pj}$$

We shall now study the variations of these parameters for pressures no higher than a few bars absolute and a temperature between 0°C and 150°C.

In these conditions, ω_j remains constant, and is independent of P and T.

The parameter $\hat{b}_k$ is assimilated to the parameter $\hat{b}_{NaCl}$, whose variation as a function of the temperature is given by the author's Figure 163A. This parameter decreases as the temperature increases.

The parameter b_{pj} expresses the short-range interactions between ions with opposing charges. We then agree that:

$$b_{pj} = b_{Na^+Cl^-} - 0.19(z_j - 1)$$

The value of $b_{Na^+Cl^-}$ at 25°C is –0.049, and it is an increasing function of temperature, shown by Figure 163B.

As the values of b_k and b_{pj} are the same for all electrolytes (the authors' Tables 29 and 30), the activity coefficient γ_j is given by:

$$\log_{10} \gamma_j = -\frac{A_\gamma z_j^2 I^{1/2}}{\Delta} + \Gamma_j + \omega_j^{abs} b_k I + b_{pj} I$$

NOTE.– Meissner's method contains only one parameter q, whereas Helgeson's contains three. Thus, we can legitimately suppose that the latter method is more precise. In reality, Meissner's method is accurate for a low ionic strength ($I < 0.5$ kmol.m^{-3}).

3.5. Osmotic coefficient (electrolytes), [MEI 80] and [HEL 81]

3.5.1. *Osmotic pressure*

Consider two recipients filled with pure water, separated by a membrane that is permeable to water but not to any solutes. The initial pressure in the two recipients is P_{A0}.

We then gradually dissolve a soluble product in recipient A. We see that water flows from recipient B to recipient A to dilute the solution therein. In other words, the pressure P_B is greater than the pressure P_A, which *has been reduced by a certain value Π*, which is the osmotic pressure (or rather the osmotic "depression" from the point of view of the solvent, as we shall see).

More specifically, the *hydrochemical* potential of water in B is μ_B and the potential of water in A is:

$$\mu_A = \mu_o + P_A V_e + RT \, Ln \, a_e$$

$$\mu_B = \mu_o + P_{A0} V_e$$

where V_e is the molar volume of water: m^3.kmol^{-1}.

If we increase the pressure P_{A0} in recipient A so that there is equilibrium between the two faces of the membrane, we should have:

$$\mu_A = \mu_B \text{ or indeed } P_A - P_{A0} = -\frac{RT \, Ln \, a_e}{V_e}$$

The osmotic pressure is, by definition:

$$\Pi = P_A - P_{A0} = -\frac{RT \, Ln \, a_e}{V_e} \text{ (generally } \Pi > 0 \text{, because } a_e < 1) \qquad [3.8]$$

In general, the activity of water $a_e \rightarrow a_w$ is less than 1, so the osmotic pressure as defined by equation [3.8] is positive. This means that the pressure P_A of the solution must be greater than the pressure P_B of pure water if we want to prevent the pure water from diluting the solution.

Nabetami *et al.* [NAB 90] describe an osmometer used to determine the osmotic depression of a soil (macromolecular dispersion).

Let us emphasize the fact that the osmotic pressure is not a direct property of the solute. This is a property of water in the presence of solute. It is incorrect to speak of the osmotic depression of a solute.

NOTE.– In the expression of Π, let us make the following changes.

$$a_e = \gamma_e x_e \neq x_e = 1 - x_s \text{ hence } Ln a_e \neq -x_s = -\frac{c_s}{c_T} \neq \frac{-c_s}{c_e}$$

However:

$$V_e c_e = \phi_e \#1$$

This gives us van't Hoff's "law":

$$\Pi = RT c_s$$

NOTE.– Consider solutes of large size (colloids, proteins, organic debris). At a notable mass fraction, their molar concentration will remain low. Consequently, the corresponding osmotic pressure can be overlooked.

3.5.2. *Osmotic coefficient*

The osmotic coefficient ϕ is such that the osmotic pressure is written:

$$\Pi = RT m_T \phi$$

where m_T is the total molality of the solution: mole per kg of water.

Indeed, the osmotic coefficient pertains only to aqueous solutions.

If dissociated electrolytes are present, the ions resulting from it are counted independently. They are characterized by the index j.

$$m_T = \sum_k v_{kj} m_k + \sum_p m_p$$

The index k characterizes the dissociated electrolytes and the index p characterizes the non-dissociated species. The numbers v_{kj} are the stoichiometric coefficients of the dissociation.

The osmotic coefficient may take a variety of forms.

1) In light of the expression of the osmotic pressure, we can write:

$$-\frac{RT}{V_e} Lna_e = RTm_T\phi$$

Hence:

$$\phi = -\frac{Lna_e}{V_e m_T}$$

However:

$$\frac{1}{V_e} = 55.51 \ kmol.m^{-3}$$

Hence:

$$\phi = -55.51 \frac{Lna_e}{m_T} \qquad [3.9]$$

2) When the solution is dilute, the activity coefficient of water is near to 1, and we have:

$$Ln\, a_e \# Ln\, x_e$$

Yet:

$$x_e \# 1 - m_T V_e = 1 - \frac{m_T}{55.51}$$

$$Lna_e \# - m_T/55.51$$

The osmotic pressure becomes (for $T = 298.15$ K):

$$\Pi = RTm_T = 24.78 \times 10^5 . m_T Pa = 24.78\ m_T bar$$

We finally see that the osmotic coefficient in equation [3.9] expresses the fact that the solution is not diluted.

3.5.3. *Calculation according to Helgeson et al. [HEL 81]*

The osmotic coefficient is given by equation 190:

$$\phi = -2.303 \left[\frac{A_\gamma I^{1/2} \sigma \Sigma_j m_j z_j^2}{3}\right] + \frac{Ln(1+0.0180153 \times m_T)}{0.0180153 m_T}$$

$$+ \frac{2.303}{2m_T} \Sigma_j m_j \left(\omega_j \Sigma_k b_k y_k I + \Sigma_p b_{jp} m_p\right)$$

We set:

$$\sigma = \mathring{a} B_\gamma I^{1/2}$$

These three parameters are defined in relation to the activity coefficients. The parameter σ is then given by:

$$\sigma = \frac{3}{\alpha^3}\left(1 + \alpha - \frac{1}{(1+\alpha)} - 2Ln(1+\alpha)\right)$$

This parameter is the same as Debye and Hückel's parameter σ, with the restriction that $\mathring{a}$ here is a mean value taken over all the ions of the solution and does not refer to any one ion in particular.

We have shown, as regards the calculation of the activity coefficients, that:

$$\Sigma_k b_k y_k I = \Sigma_k \left[b_k m_k \Sigma_j \left(\frac{v_{jk} z_j^2}{2}\right)\right]$$

ω_j is given by the authors' Table 3 and is expressed in cal.mol^{-1}.

The coefficient b_k is given by the authors' Table 4 and is expressed in L.cal^{-1}.

The coefficient b_{jp} is given by the authors' Table 7.

m_T is the total ionic molality (concentration):

$$m_T = \sum_j m_j$$

EXAMPLE 3.7.–

Let us find the osmotic coefficient of a solution of NaCl, in which I = 3, at 25°C.

$$A_\gamma = 0.5091 \quad \omega_{Cl^-} = 1.4560 \times 10^5 \quad b_k = 1.7865 \times 10^{-6}$$

$$B_\gamma = 0.3283 \times 10^{10} \quad \omega_{Na^+} = 0.3305 \times 10^5 \quad b_{pj} = 0.096$$

$$å = 3.72\,Å \quad I = 3 = m_{NaCl} = m_k \quad m_T = 6 = m_{Na^+} + m_{Cl^-}$$

$$y_k I = \psi_k m_k = \sum_j \mu_{jk}\, m_k \frac{z_j^2}{2}$$

For NaCl:

$$\nu_{jk} = 1 \quad m_k = 3 \quad z_j = 1$$

$$\sum_k b_k\, y_k\, I = b_k\, m_k = 1.7865 \times 10^{-6} \times 3 = 5.3595 \times 10^{-6}$$

$$\alpha = 3.72 \times 10^{-10} \times 0.3289 \times 10^{10} \times 3^{1/2} = 2.119178$$

$$\sigma = \frac{3}{2.119178^3}[3.119178 - 0.320597 - 2.275139]$$

$$\sigma = 0.165$$

$$\phi = -\frac{1.172457 \times 3^{1/2} \times 0.165 \times 3 \times 2}{3} + \frac{Ln\left(1 + 0.0180153 \times 6\right)}{0.0180153 \times 6}$$

$$+ \frac{2.303}{2 \times 6}\left[\left(1.4560 \times 10^5 + 0.3305 \times 10^5\right) \times 5.3595 \times 10^{-6} \times 3 + 2 \times 3 \times 0.096 \times 3\right]$$

$$\phi = 1.162$$

3.5.4. *Meissner's calculation*

Knowing the ionic strength of the solution and the parameter q of the dissolved electrolyte, the lattice of curves in Figure 2 of [MEI 80] gives the value of the activity de water. The osmotic coefficient ϕ and the osmotic pressure Π can be deduced from this, as follows:

$$\phi = \frac{-55.51 \, Lna_e}{m_T} \quad \text{and} \quad \Pi = -\frac{RTLna_e}{V_e}$$

EXAMPLE 3.8.–

Solution of NaCl for which I = 3 at 25°C (m_T = 6).

For this salt:

q = 2.23 [KUS 78]

On the lattice of curves, we read [MEI 80]:

a_e = 0.89

Hence:

$$\phi = -\frac{55.51 Ln\,0.89}{6} = 1.08$$

The value is near to the value of 1.16 given by Helgeson's method. We shall employ Helgeson's value, because Meissner's method may exhibit an error of up to 20%, sometimes, for values of the ionic strength greater than 1.

In addition, Meissner's method uses only one parameter, whilst Helgeson's uses three.

3.6. Cruz and Renon's theoretical and practical review [CRU 78]

3.6.1. *Debye and Hückel's term [DEB 23]*

Cruz and Renon note that Debye and Hückel make an improper simplification. Indeed, the latter two authors write:

$$V / \overline{v}_o = n_o$$

V: total volume of the solution

$\overline{v}_o$: molar volume of the solvent

$\overline{v}_i$: partial molar volume of an ion: $cm^3.ion\text{-}mole^{-1}$

$\overline{v}_j$: partial molar volume of a molecule: $cm^3.mole^{-1}$

n_o: number of molecules of solvent

In reality, Debye and Hückel ought to have written:

$$V / v_o = n_o + \sum_{i=1} \frac{n_i \overline{v}_i}{v_o} + \sum_{j=1} \frac{n_j \overline{v}_j}{v_o}$$

Also, Debye and Hückel's coefficient χ_i is replaced by the fraction $1/(1 + K\mathring{a})$.

With the restrictions satisfied, the expression given by Cruz and Renon is perfectly equivalent to that of Debye and Hückel with regard to $Ln\gamma_i$.

Cruz and Renon, like Debye and Hückel, make the partial molar volumes identical to the molar volumes in the pure state:

$$\overline{v}_j = v_j \text{ and } \overline{v}_i = v_i$$

The volumes of the ions do not directly play a role, because they are supposed to belong to non-dissociated molecules.

3.6.2. *Born's equation [BOR 20]*

According to Durand [DUR 53], we can define:

– the electrical field by $\vec{E}$, and

– the electrical induction by $\vec{D}$.

We suppose the medium to be isotropic and, in this case, E and D are collinear.

$$\vec{D} = \varepsilon_r \varepsilon_o \vec{E}$$

ε_0: permittivity of vacuum;

ε_r: relative permittivity of the medium in question.

The volumetric energy density is:

$$\frac{1}{2}\vec{E}\vec{D} = \frac{1}{2\varepsilon_0\varepsilon_r}D^2 = \frac{\varepsilon_r\varepsilon_0}{2}E^2$$

However, the field outside of a ball with charge z_e and radius r_i is:

$$E = \frac{ze}{4\pi\varepsilon\varepsilon_o r^2} \quad r > r_i \ (S.I.)$$

Let us integrate over a solid angle of 4π and with r_i infinity. The elementary volume is:

$$d\Omega = 4\pi r^2\, dr$$

The total spatial energy is expressed by:

$$W = \frac{4\pi\varepsilon_r\varepsilon_0 z^2 e^2}{2 \times 16\pi^2 \varepsilon_r^2 \varepsilon_0^2}\int \frac{r^2}{r^4}r^2 = \frac{z^2 e^2}{8\pi\varepsilon_1\varepsilon_0}\int_{r_i}^{\infty}\frac{d^2}{r^2} = \frac{z^2 e^2}{8\pi\varepsilon_r\varepsilon_0 r_i}$$

When the same ion transits from the very dilute (pure) solvent into a solution, the total spatial energy increases by:

$$\frac{z^2 e^2}{8\pi r_i \varepsilon_o}\left(\frac{1}{\varepsilon_r}-\frac{1}{\varepsilon_{ro}}\right) \text{(S.I.)}$$

ε_{ro} : relative permittivity of the pure solvent,

ε_r : relative permittivity of the solution,

and, for an electrolyte, the excess Helmholtz energy of the medium corresponding to dissolution is (for a kilomole):

$$F_{Born}^E = \frac{e^2}{8\pi\varepsilon_o}\left(\frac{1}{\varepsilon_r}-\frac{1}{\varepsilon_{ro}}\right)\left(\frac{v_A z_A^2}{r_A}+\frac{v_B z_B^2}{r_B}\right)$$

r_A and r_B are the radii of the ions A and B in the solution.

z_A and z_B are the charges of those ions.

v_A and v_B are the stoichiometric coefficients of the electrolyte $A_{vA}B_{vB}$.

3.6.3. NRTL term

The NRTL equation uses the parameters:

$$G_{ki} = \exp\left(-\alpha_{ki}\frac{C_{ri}}{RT}\right) \text{ and } Z_{ki} = G_{ki}C_{ki} \text{ and } C_{ii} = 0$$

However, in the solution of a weak electrolyte AB in a solvent S, we find ourselves with four chemical species.

AB, A, B and S.

Cruz and Renon conserve only the following couples [k,i]:

$$[A,S]; [B,S]; [AB,S] \text{ and } [S,AB]$$

The parameters G_{ki} of the other couples are null. Thus:

$$[S, A]; [B, B]; [A, B] \text{ etc....}$$

However, the excess Helmholtz energy is very generally written:

$$F_{NRTL}^E = \sum_i n_i \frac{\sum\limits_k n_k Z_{ki}}{\sum\limits_j n_j G_{ji}}$$

which, with the above restrictions in place, becomes:

$$F_{NRTL}^E = n_{AB} \left[\frac{n_S Z_{S,AB}}{n_S G_{S,AB} + n_{AB}} \right] + n_S \left[\frac{n_A Z_{AS} + n_B Z_{BS} + n_{AB} Z_{AB,S}}{n_A G_{AS} + n_B G_{BS} + n_{AB} G_{AB,S} + n_S} \right]$$

$$-n_A Z_{A,S} - n_B Z_{B,S} \quad \text{where} \quad n_A / v_A = n_B / v_B$$

In reality, the NRTL relation applies to the excess Gibbs energy rather than the excess Helmholtz energy, but it is accepted that all considerations regarding electrolytes *take place at constant volume*.

CONCLUSION.–

Cruz and Renon's analytical expression pertains to the activity coefficients of weak electrolytes, i.e.:

– of non-dissociated electrolyte;

– of ions of dissociated electrolyte;

– of solvent.

A fortiori, this expression is valid for strong (fully dissociated) electrolytes – salts.

Acids are never totally dissociated, however strong they may be. With regard to polyacids, it is generally enough to take account of the first dissociation constant – that which releases a single proton (with the exception of sulfuric acid).

It must be borne in mind that any liquid–vapor equilibrium is established between the vapor phase and the *non-dissociated electrolyte present in the solution*. The vapor pressure of the ions is very usually negligible (such is the case, in particular, with salts, which produce ionic crystals).

The equilibrium constant for the dissociation of the electrolyte $A_{\nu_A} B_{\nu_B}$ is defined by:

$$K(T) = \frac{\left(x_A \gamma_A\right)^{\nu_A} \left(x_B \gamma_B\right)^{\nu_B}}{x_{AB} \gamma_{AB}}$$

x_A, x_B and x_{AB} are the true molar fractions of A, B and AB. In the next equation, m_{AB} is the molality of the electrolyte that is supposed to be non-dissociated, and α is the dissociated fraction.

$$K_m(T) = \frac{\gamma_\pm^\nu \alpha^\nu \nu_A^{\nu_A} \nu_B^{\nu_B} m_{AB}^\nu}{\gamma_{AB} m_{AB}(1-\alpha)} \quad \left(\text{where } \nu = \nu_A + \nu_B\right)$$

Cruz and Renon [CRU 79] examined the parameters of aqueous solutions of a weak acid: acetic acid.

Finally, the work of Cruz and Renon has the interest of giving the calculation parameters regarding mixtures of water with:

– 7 acids (HCl, HBr, HI, $HClO_4$, HNO_3, H_2SO_4, H_3PO_4);

– caustic soda;

– 7 salts (KCl, NaCl, LiCl, LiBr, $CaCl_2$, $CaBr_2$, NH_4NO_3).

Laws Governing the Transfer of Material and Heat Between Two Immiscible Fluids Electrolytes and their Diffusion

4.1. Fick's law [FIC 55]

4.1.1. *Expression of Fick's law*

This law expresses the movement of a single component in an immobile environment. Strictly speaking, this situation can only arise if the immobile medium is a solid. In this case, the migrating species only experiences friction with the fixed medium and, at constant pressure and temperature, we can write:

$$-c\frac{d\mu}{dz} = \frac{RT}{D}N$$

N: absolute flux density of the migrating species, i.e. relative to the workshop: kmol.m^{-2}.s^{-1}. These flux densities *are positive in the direction of increasing z values*;

– c: number of kilomoles of the migrating species in the unit volume of the fixed medium: kmol.m^{-3};

– D: diffusivity of the migrating species in the immobile medium: m^2.s^{-1};

– T: absolute temperature: K;

– R: perfect gas constant: 8314 J.kmol^{-1}.K^{-1};

– μ: chemical potential of the migrating species: $J.kmol^{-1}$;

– z: dimension parallel to the motion of the migrating species: m

An excellent example where Fick's law applies is the migration of humidity during the air drying of a piece of wood. That drying process may take several months. The migration of certain products in gels or cheese also obeys Fick's law. Finally, this law is also used in the study of two-dimensional diffusion on the surface of crystals. In fact, this law is very frequently employed in physics, rightly or wrongly.

As the medium is immobile, the chemical potential depends only on the concentration of the migrating species. Let us set:

$$\mu = RTLnc + const. \text{ Hence } \frac{d\mu}{dz} = \frac{RT}{c}\frac{dc}{dz}$$

We obtain the usual expression of Fick's law:

$$N = -D\frac{dc}{dz}$$

Fick's law, remember, imposes that the species transported is unique. However, in liquids and gases, a species never migrates on its own, and thus it is no longer possible to consider the medium to be immobile. Except for dissociated electrolytes, then, we need to use the Maxwell–Stefan law, which takes account of the interactions between molecules of different species.

4.1.2. A single species transferred between two fluids not transferred (elementary two-film theory)

Let us give three different expressions for the molar flux density which crosses the interface:

$$\beta_A(x_{AI} - x_A) = \beta_B(x_B - x_{BI}) = K_A(x_A^* - x_A)$$

K_A : global transfer coefficient in relation to the phase A: $kmol.m^{-2}.s^{-1}$

x_A^* : molar titer in the phase A at equilibrium with the phase B whose titer is x_B

In the first equation, the left-hand side characterizes the film on the side of the phase A and the right-hand side the film on the side of the phase B.

By locally assimilating the equilibrium curve to a straight line, we can write:

$$x_A = mx_B^* + x_0 \qquad\qquad [4.1]$$

$$x_A^* = mx_B + x_0 \qquad\qquad [4.2]$$

$$x_{AI} = mx_{BI} + x_0 \text{ (equilibrium at the interface)}$$

$$x_A^* - x_{AI} = m(x_B - x_{BI})$$

Thus:

m: mean slope of the equilibrium curve.

We can write the following identities:

$$x_A^* - x_A = (x_A^* - x_{AI}) + (x_{AI} - x_A) = m(x_B - x_{BI}) + (x_{AI} - x_A)$$

Hence:

$$\frac{x_A^* - x_A}{\dfrac{1}{K_A}} = \frac{x_{AI} - x_A}{\dfrac{1}{\beta_A}} = \frac{x_B - x_{BI}}{\dfrac{1}{\beta_B}} = \frac{m(x_B - x_{BI})}{\dfrac{m}{\beta_B}} = \frac{x_A^* - x_{AI}}{\dfrac{m}{\beta_B}} = \frac{x_A^* - x_A}{\dfrac{m}{\beta_B} + \dfrac{1}{\beta_A}}$$

Finally, for the global transfer coefficient K_A, in relation to the phase A:

$$\frac{1}{K_A} = \frac{1}{\beta_A} + \frac{m}{\beta_B} \qquad\qquad [4.3]$$

As we have knowledge of the equilibrium curve, either in graphical or analytical form, the given values of x_A and x_B can be used to obtain x_B^* and x_A^*. Hence, in light of equations [4.1] and [4.2], we obtain:

$$m = \frac{x_A - x_A^*}{x_B^* - x_B} \text{ and } x_0 = \frac{x_A x_B - x_A^* x_B^*}{x_B - x_B^*}$$

The parameter m is the slope of the string linking points A and B on the equilibrium curve.

The classic two-film theory has been validated by experience.

NOTE.– We can apply the above method to a complex mixture such as those found in distillation. The liquid–vapor equilibrium is written as follows for the component i:

$$\phi_i y_i P_T = x_i \gamma_i \pi_i(t)$$

t: temperature: °C

Differentiate:

$$P_T\, dy_i \left(\phi_i + y_i \frac{\partial \phi_i}{\partial y_i} \right) = \pi_i(t)\, dx_i \left(\gamma_i + x_i \frac{\partial \gamma_i}{\partial x_i} \right)$$

Hence:

$$m_i = \frac{dy_i}{dx_i} = \frac{\pi_i(t)\left(\gamma_i + x_i \dfrac{\partial \gamma_i}{\partial x_i} \right)}{P_T\left(\phi_i + y_i \dfrac{\partial \phi_i}{\partial y_i} \right)}$$

Similarly, for solvent-based extraction, the equilibrium is written thus, after differentiation:

$$m_i = \frac{dx_{li}}{dx_{2i}} = \frac{\gamma_{2i} + x_{2i} \dfrac{\partial \gamma_{2i}}{\partial x_{2i}}}{\gamma_{li} + x_{li} \dfrac{\partial \gamma_{li}}{\partial x_{li}}}$$

4.2. Maxwell–Stefan equations

4.2.1. Expression of the Maxwell–Stefan equations

Slattery [SLA 72] gives a presentation of the Maxwell-Stefan (M.–S) equations.

As with Fick's law, the macroscopic movement of each species takes place in the direction corresponding to a decrease in its chemical potential. This direction corresponds to $-d\mu_i/dz$, with the fluxes being positive in the direction of increasing z.

During its displacement, each molecule of type i exchanges energy with the molecules j. That energy is proportional to the difference $(v_i - v_j)$ between their velocities. It is also proportional to the molar fraction of the molecules j. In a fluid mixture with n components and for the species i, the M.–S equation is written:

$$-\frac{d\mu_i}{dz} = \sum_{j=1}^{n} f_{ij} x_j (v_i - v_j)$$

$$j \neq i$$

Here, f_{ij} is a "friction coefficient" (not dimensionless).

To express the transfer of material in the permanent regime between two fluids, we shall use the frame of reference for which the profiles of the chemical potentials are immobile. That reference is that of the workshop.

We can posit:

$$v_i = N_i / c_i \qquad v_j = N_j / c_j \qquad f_{ij} = \frac{RT}{D_{ij}} \quad \frac{1}{D_{ii}} = 0$$

The equation of diffusion becomes:

$$-\frac{1}{RT}\frac{d\mu_i}{dz} = \sum_{j=1}^{n} x_j \left[\frac{N_i}{c_i} - \frac{N_j}{c_j} \right] \frac{1}{D_{ij}}$$

However, the traditional presentation of this equation is different. Let us set:

$$c_T = \sum_{i=1}^{n} c_i \qquad c_i = x_i c_T \qquad c_j = x_j c_T$$

By multiplying both sides of this equation by c_i, we obtain:

$$-\frac{c_i}{RT}\frac{d\mu_i}{dz} = \sum_{j=1}^{n}\left[\frac{N_i x_j - N_j x_i}{D_{ij}}\right]$$

4.2.2. Symmetry of the diffusivities

For a mixture with n components, we accept that:

$$D_{ij} = D_{ji}$$

This symmetry remains compatible with any dependence of the D_{ij} values on the local composition, as we shall see later on.

4.2.3. Complementary equation

At constant T and P, let us add together, member by member, the n M.–S equations, bearing the Gibbs–Duhem equation in mind:

$$\sum_{i=1}^{n}\frac{c_i}{RT}\frac{\Delta\mu_i}{e} = 0 = \sum_{i=1}^{n}\sum_{j=1}^{n}\frac{N_i x_j}{D_{ij}} - \sum_{i=1}^{n}\sum_{j=1}^{n}\frac{N_j x_i}{D_{ij}} = 0$$

Thus, the sum of the n M.–S equations is expressed by an identity:

$$0 = 0$$

These equations, therefore, are not independent, and the calculation of the n flux densities N_i will require a *complementary equation*.

We shall separately examine diffusion in a liquid medium and diffusion in a gaseous medium.

1) Diffusion in a liquid

Consider two identical recipients A and B, juxtaposed and linked at the bottom by a horizontal tube of moderate diameter, with a tap. The two recipients are filled with mixtures of the same nature, but different compositions. The levels of filling are such that the hydrostatic pressure is the same at both ends of the tube. When the tap is opened, diffusional transport begins to take place. Let us set:

S: section of the tube: m^2

N_i: flux density of the component i oriented positively from B towards A: $kmol.m^{-2}.s^{-1}$

M_i: molar mass of the component i: $kg.kmol^{-1}$

In order for the hydrostatic equilibrium to be maintained, it is necessary for the overall mass transferred from one recipient to the other to be zero.

$$S\sum_i N_i M_i = 0 \text{ or, quite simply } \sum_i N_i M_i = 0$$

This last equation is the complementary equation we are looking for.

2) Diffusion in a gas

The aforementioned two recipients have identical volumes and are filled with gaseous mixtures of the same components, but at different concentrations. Initially, the pressures are supposed to be identical in both recipients. When the tap is opened and if we accept the perfect gas law, the number of kilomoles present in each recipient must not vary in order for the pressures to remain equal. Thus:

$$\sum_i N_i = 0$$

We have just written the complementary equation relative to gases.

These two cases of diffusion are supposed to take place in a *single phase*.

3) Distillation.

In this operation, there is no essential component. Indeed, all the components are transferred. However, it is possible to perform an approximate calculation by having the complementary equation express the hypothesis that the heat released by the condensation of the vapor is entirely used up to vaporize the liquid. The complementary equation then becomes:

$$\sum_{i=1}^{n} \Lambda_i N_i = 0$$

Λ_i: latent heat of state change: $J.kmol^{-1}$.

Some authors have proposed writing, still for distillation:

$$\sum_{i=1}^{n} N_i = 0$$

This complementary equation is manifestly wrong.

4.2.4. Canonical M.–S equation

The average diffusivity $\overline{D}_1$ for the component i is obtained by:

$$\frac{1}{\overline{D}_i} = \sum_{j=1}^{n} \frac{x_j}{D_{ij}}$$

We then calculate *all* the flux densities N_i and N_j by the elementary double film method. This method is founded on Fick's law:

$$-c_i \frac{d\mu_i}{dz} = \frac{RT}{\overline{D}_i} N_i$$

We then define the equivalent flux density of entrainment N_{Ei} for the component i by:

$$\frac{N_{Ei}}{\overline{D}_i} = \sum_{\substack{j=1 \\ j\neq i}}^{n} \frac{N_j}{D_{ij}}$$

The M.–S equation of the component i is then written:

$$-c_i \frac{d\mu_i}{dx} = \frac{RT}{\overline{D}_i} \left(N_i - x_i N_{Ei} \right)$$

or indeed:

$$N_i = -\frac{\overline{D_i}}{RT} c_i \frac{d\mu_i}{dx} + x_i N_{Ei}$$

We use the term "canonical equation" to speak of this form of the M.–S equation.

On the right-hand side of this equation:

– the first term is Fick's term;

– the second term is the entrainment term.

4.2.5. *Two immiscible fluid phases*

We shall stop looking for a complementary equation that has the appearance of a homogeneous linear form relative to the flux densities N_i.

We posit:

$$N_i^* = -\frac{\overline{D_i}}{RT} c_i \frac{d\mu_i}{dz} \quad \text{and} \quad N_j^* = -\frac{\overline{D_j}}{RT} c_j \frac{d\mu_j}{dz}$$

The canonical equation is then written:

$$N_i = N_i^* + x_i N_{Ei}^*$$

The N_{Ei}^* are calculated on the basis of the N_j^* and the D_{ij} of each phase. Thus:

– the N_i^* will have been calculated by the two-film theory (see section 4.1.2) and will not vary from one phase to another;

– the N_{Ei} will be slightly different from one phase to the next because the D_{ij} will be different between the α phase and the β phase.

Thus, we shall simply write:

$$N_{Ei}^* = \sum_{\substack{j=1 \\ j \neq i}}^{n} N_i^*$$

and, finally:

$$N_i = N_i^* + \eta_i x_i N_{Ei}^*$$

The coefficient η_i is what could be called the "entrainment efficiency". It is between 0 and 1, and probably less than 0.3, and needs to be determined experimentally.

The value of x_i is taken at the interface on the upstream side in relation to N_{Ei}. The term $x_i N_{Ei}^*$ is a sort of convection term.

NOTE.– The aforementioned may help to understand a possible displacement of the interface in a tube.

4.2.6. *Important specific cases*

Hereafter, we shall suppose that the two phases (whether miscible or otherwise) are nearly isotonic, meaning that the activity of the two solvents (immiscible phases) or of the single solvent (miscible phases) has a similar value in both phases. Hence, there is practically no solvent transfer from one phase to the other.

Now suppose that, of the solutes, only one has a noteworthy concentration in the two phases. Let X be the order of magnitude of these two concentrations. This solute is known as the essential solute.

Suppose, furthermore, that the other solutes appear only at low concentrations in both phases and let ε be the order of magnitude of those concentrations. These solutes are called the secondary solutes. The flux densities of those minor solutes are around ε.

As regards the essential solute, the Fickian term in the canonical equation will be of the order of X and the order of magnitude of the entrainment term will be the same as the product $(X \times \varepsilon)$ – i.e. less than X. In other words, the Fickian term will be predominant over the entrainment term.

Therefore, whether it is a case of the stripping of a *single* gas from a liquid by an inert gas, or the solvent-based extraction of a *single* compound, Fick's law will apply in each phase and we can use the elementary two-film theory with no problems. The same will be true with the absorption of a *single* gas from a gaseous mixture into a liquid.

On the other hand, in the case of secondary solutes, entrainment may have an impact on the transfer of those solutes.

NOTE.– When a liquid is filtered on a membrane for all solutes, it will be the entrainment term which is predominant, but we shall see later on that the problem is dealt with differently (see Chapter 3).

4.3. Matrix resolution of the M.–S equations

4.3.1. *Matrix expression of the gradient of chemical potential*

All the molar fractions are variable, so that:

$$\frac{d\mu_i}{dz} = \sum_{j=1}^{n-1} \frac{\partial \mu_i}{\partial x_j} \frac{dx_j}{dz} \qquad i = 1, 2, ..., n-1$$

However:

$$\frac{\partial \mu_i}{\partial x_j} = RT \frac{\partial \ln \gamma_i x_i}{\partial x_j} = RT \left[\frac{\partial \ln x_i}{\partial x_j} + \frac{\partial \ln \gamma_i}{\partial x_j} \right]$$

$$= \frac{RT}{x_i} \left[\delta_{ij} + \frac{x_i}{x_j} \frac{\partial \ln \gamma_i}{\partial x_j} \right] = \frac{RT}{x_i} \Gamma_{ij}$$

Hence:

$$x_i \frac{d\mu_i}{dz} = \sum_{j-1}^{n-1} \Gamma_{ij} \frac{dx_j}{dz} \qquad i = 1, 2, ..., n-1$$

and, in matrix notation:

$$\left(x \frac{d\mu}{dz} \right) = RT [\Gamma] \left(\frac{dx}{dz} \right)$$

4.3.2. *Matrix expression of the M.–S equations*

For the component i, the M.–S equation is written:

$$\frac{x_i}{RT}\frac{d\mu_i}{dz} = -N_i\sum_{\substack{k=1\\k\neq i}}^{n}\frac{x_k}{c_T D_{ik}} + x_i\sum_{\substack{ki\\k\neq i}}^{n-1}\frac{N_k}{c_T D_{ik}} + x_i\frac{N_n}{c_T D_{in}}$$

However, if we accept that the complementary equation is linear and homogeneous:

$$N_n = -\sum_{k=1}^{n-1}a_k N_k = -\sum_{\substack{ki\\k\neq i}}^{n-1}a_k N_k - a_i N_i$$

Hence:

$$\frac{x_i}{RT}\frac{d\mu_i}{dz} = -N_i\left[\sum_{\substack{k=1\\k\neq i}}^{n}\frac{x_k}{c_T D_{ik}} + \frac{x_i a_i}{c_T D_{in}}\right] + \sum_{\substack{j=1\\K\neq i}}^{n-1}\left[\frac{1}{c_T D_{ik}} - \frac{a_k}{c_T D_{nk}}\right]N_k \, x_i$$

Let us posit:

$$z = \eta e \qquad 0 \leq \eta \leq 1$$

e: thickness of the limiting film bordering the phase in question: m

$$\eta = 0 \qquad x_i = x_{i\infty}$$

$$\eta = 1 \qquad x_i = x_{il}$$

The sign ∞ characterizes the mass of the phase (the bulk of the phase) and the sign I characterizes the interface with another phase.

Let us also set:

$$\beta_{ij} = c_T\frac{D_{ij}}{e}$$

β_{ij}:   binary transfer coefficient between i and j: $kmol.m^{-2}.s^{-1}$

The M.–S equation is then written as follows (if we replace the indices k with the index j):

$$\frac{x_i}{RT}\frac{d\mu_i}{d\eta} = -N_i \left[\sum_{\substack{j=1\\j\neq 1}}^{n-1}\frac{x_j}{\beta_{ij}} + \frac{x_i a_i}{\beta_{in}}\right] + \sum_{\substack{j=1\\j\neq 1}}^{n-1}\left[\frac{1}{\beta_{ij}}\frac{a_k}{\beta_{nj}}\right]N_j x_i$$

Finally, let us introduce a matrix $[B]$ defined by:

$$B_{ii} = \frac{x_i a_i}{\beta_{in}} + \sum_{\substack{j=1\\j\neq i}}^{n}\frac{x_j}{\beta_{ij}} \quad \text{and} \quad B_{ij} = x_i\left[\frac{a_j}{\beta_{nj}} - \frac{1}{\beta_{ij}}\right]$$

Therefore, in matrix notations:

$$\frac{1}{RT}\left(x\frac{d\mu}{d\eta}\right) = -[B](N) = [\Gamma]\left(\frac{dx}{d\eta}\right)$$

Hence, the value of the first (n-1) flux densities:

$$(N) = -[B]^{-1}[\Gamma]\left(\frac{dx}{d\eta}\right) = [k]\left(\frac{dx}{d\eta}\right)$$

$[B]$: matrix of inverses of binary transfer coefficients: $m^2.s.kmol.^{-1}$

When η varies from 0 to 1, we need to find the mean value of $(dx / d\eta)$.

We take:

$$\left(\overline{\frac{dx}{d\eta}}\right) = \left(\frac{x_1 - x_\infty}{1 - 0}\right)$$

Finally:

$$(N) = [k](x_1^B - x_\infty^B)$$

$[k]$: partial transfer matrix: $kmol.m^{-2}.s^{-1}$

4.3.3. *Global transfer matrix*

Consider two phases A and B whose partial transfer matrices are:

$$\left[k^A \right] \quad \text{and} \quad \left[k^B \right]$$

The matrix $\left[k^A \right]$ enables us to write:

$$\left(x_\infty^A - x_I^A \right) = \left[k^A \right]^{-1}(N) \quad \text{and} \quad \left(x_I^B - x_\infty^B \right) = [k^B]^{-1}(N) \qquad [4.4]$$

and, with the matrix $\left[k^B \right]$ regardless of $[M]$:

$$[M] \left[k^B \right]^{-1}(N) = [M](x_I^B - x_\infty^B) \qquad [4.5]$$

However, in light of the equilibrium relation, we can write (see section 4.1.2):

$$\left(x_I^A \right) = [M]\left(x_I^{B^*} \right) + (b)$$

$$\left(x_\infty^{A^*} \right) = [M](x_\infty^B) + (b)$$

We shall subtract these last two equations from one another, term by term:

$$(x_I^A - x_\infty^{A^*}) = [M](x_I^B - x_\infty^B) \qquad [4.6]$$

By substituting equation [4.4] into equation [4.6]:

$$(x_I^A - x_\infty^{A^*}) = [M]\left[k^B \right]^{-1}(N) \qquad [4.7]$$

Let us add equations [4.4] and [4.7] together, term by term:

$$(x_\infty^A - x_\infty^{A^*}) = \left[k^A \right]^{-1}(N) + [M]\left[k^B \right]^{-1}(N) = \left[K_0^A \right]^{-1}(N)$$

The matrix $\left[K_O^A \right]$ is the global transfer matrix in relation to the phase A:

$$\left[K_O^A \right]^{-1} = \left[k^A \right]^{-1} + [M]\left[k^B \right]^{-1}$$

The matrix [M] is obtained by generalizing equation [4.1]:

$$(x_A) = [M](x_B^*) + (x_O)$$

4.3.4. Calculation procedure

(N) can be calculated with no difficulty, provided we know the expression of the complementary equation. For this, we need to have the molar fractions at the interface if we use the method presented in section 4.2.3.3.

To get the calculation started, we take:

$$x_I^A = \frac{1}{2}(x_\infty^A + x^{A*}) \quad \text{and} \quad x_I^B = \frac{1}{2}(x_\infty^B + x^{B*})$$

Having obtained the transfer matrices and the flux densities, N_i, we calculate:

$$(x_I^B) = (x_\infty^B) - \left[k^A \right]^{-1} (N)$$

$$(x_I^B) = (x_\infty^B) - \left[k^B \right]^{-1} (N)$$

Thus, we find a more accurate value for the transfer matrices and the flux densities. Two or three iterations will suffice.

4.3.5. Krishna and Standart's method [KRI 79]

This method is more complicated for two reasons:

– Krishna and Standart do not calculate the N_i values directly, but instead go through the intermediary of the flux densities J_i, the sum of which is zero;

– the authors also bring into play what they call a correction matrix [Ξ] (pronounced "uppercase Xi"), which is complicated to calculate and involves exponentials of matrices. In addition, it is an iterative calculation which does not always converge on a solution.

These two complications have not truly been shown to be justified by experience, so we have not employed them here.

4.3.6. *Variation of the size of a drop*

Let the flux densities be oriented positively toward the center of the drop. We then have:

$$\sum N_i \bar{v}_i = \frac{dr_g}{d\tau} -$$

$\bar{v}_i$:   partial molar volume of the transferred products: $m^3 \cdot kmol^{-1}$

r_g :   radius of the drop: m

τ :   time: s

N_i :   molar flux density of the species i: $kmol.m^{-2}.s^{-1}$

4.3.7. *Which equation is best suited?*

Consider the example of liquid–liquid extraction.

If one species is transferred between two practically-immiscible liquids, we can agree that the transfer occurs between two stagnant fluids in relation to the interface. Fick's equation is then appropriate, as well as the elementary two-film theory.

If we now consider the ternary diagram containing a miscibility gap and if we approach the critical point where the two phases mix, then the transfer pertains to all three species present: the solvent S, the raffinate (or residue) R and, obviously, the component T which needs to be transferred. It is preferable, then, to use the M.–S equations.

Note that, because of the Gibbs–Duhem equation and the complementary equation (homogeneous in relation to the N_i values), the M.–S equations can only be used if at least two components are transferred.

4.3.8. *Transfer coefficients. Some references*

Chilton and Colburn [CHI 34] gave the expression of the heat transfer coefficients and material transfer coefficients. These relations are still used today.

Chilton and Colburn presented their results in such a way that an analogy can be drawn between the transfer coefficients for material, heat and momentum. This similarity is particularly useful in the operation of drying.

When a gas comes into contact with a liquid for a brief period of time, Higbie [HIG 35] proposes to solve the problem of transfers using what is now known as the penetration theory.

With regard to problems of fermentation, Calderbank [CAL 58, CAL 59] gave the method for calculating the gas–liquid interfacial area and the corresponding transfer coefficients.

4.3.9. *Kinetics of solvent-based extraction*

[DAN 80] looked at:

– reagents (acid, basic or neutral);

– extraction regimes (kinetic, diffusional or mixed);

– laboratory-based study techniques.

Danesi, Chiarizia and Van de Grift [DAN 80b] studied different specific cases.

4.4. Simultaneous transfers of material and heat

4.4.1. *Interfacial temperature*

Consider two fluids A and B in relative motion along a planar interface and positively evaluate the transfer flux densities N_i across the interface from phase A to at temperature t_A to phase B at temperature t_B.

As they pass from fluid A to fluid B, the N_i kilomoles of component i absorb heat $N_i(H_{Bi} - H_{Ai})$, which is taken from both of the fluids – i.e. drawn from the environment of the interface.

It is equivalent to say that the *opposite* of this heat is gained by the environment, meaning that it is released by the transfer. For the set of components transferred at the interface, we obtain:

$$Q_I = \sum_i N_i (H_{Ai} - H_{Bi}) = \sum_i N_i [C_{Ai}(t_A - t_I) + \Lambda_{ABi} + C_{Bi}(t_I - t_B)] \quad [4.8]$$

H_{Xi}: partial enthalpy of component i in the fluid X: $J.kmol^{-1}$;

Λ_{ABi}: latent heat released at the interface for the transfer of component i from A to B: $J.kmol^{-1}$.

For liquefaction, Λ_{ABi} is positive:

C_{Xi}: partial specific heat capacity of component i in the phase X: $J.kmol^{-1}.°C^{-1}$;

N_i: flux density of component i, counted positively from A to B: $kmol.m^2.s^{-1}$.

The heat Q_I is distributed between the two fluids by conduction across the two limiting films:

$$Q_I = \alpha_A (t_I - t_A) + \alpha_B (t_I - t_B)$$

Hence:

$$t_I = \frac{Q_I + \alpha_A t_A + \alpha_B t_B}{\alpha_A + \alpha_B} \quad [4.9]$$

If we eliminate Q_I between equations [4.8] and [4.9], for a gas and a liquid, we obtain:

$$t_I = \frac{1}{A}(t_o + B_G t_G + B_L t_L)$$

$$A = 1 - \frac{\sum_i N_i (C_{Li} - C_{Gi})}{\alpha_G + \alpha_L} \qquad t_o = \frac{\sum_i N_i \Lambda_i}{\alpha_G + \alpha_L}$$

$$B_G = \frac{1}{\alpha_G + \alpha_L}(\alpha_G + \sum_i N_i C_{Gi}) \qquad B_L = \frac{1}{\alpha_G + \alpha_L}(\alpha_L - \sum_i N_i C_{Li})$$

Here, the N_i values are counted positively from G to L.

NOTE.– In gas–liquid packed beds, the coefficient α_G is much smaller than the coefficient α_L, and we sometimes write:

$$Q_I = \alpha_L (t_I - t_L)$$

All of the heat released at the interface during the absorption of a gas into a liquid is then supposed to be absorbed by the liquid.

4.4.2. Evolution of temperatures of the fluids

Consider a bed where an exchange of material takes place between a liquid and a gaseous phase. We define an elementary zone, as follows:

– for a tray column, the elementary zone shall be the volume produced by the section of the column by the distance between two consecutive trays;

– for a packed column, the elementary zone shall be the volume produced by the section of the column by the thickness of an arbitrarily-chosen "slice", with that thickness being less than or equal to a quarter or a fifth of the packed height.

The thermal power received by fluid A in an elementary zone of volume ΔV will be:

$$\Delta Q_A = \alpha_A (t_I - t_A) \Delta V a_e \qquad\qquad [4.10]$$

a_e: effective area for the transfer per unit volume of the device: m^{-1}

This power causes a variation in the temperature t_A and the corresponding balance is written:

$$\Delta Q_A = \left[(W_A + \Delta W_A)(t_A + \Delta t_A) - W_A t_A \right] \ C_A \qquad [4.11]$$

W_A : flowrate of the fluid A: $kmol.s^{-1}$;

C_A : partial specific heat capacity of component k in fluid A: $J.kmol^{-1}{}^{\circ}C^{-1}$.

According to equation [4.11]:

$$\Delta t_A = \frac{(\Delta Q_A / C_A) - \Delta W_A t_A}{W_A + \Delta W_A}$$

Also:

$$\Delta W_A = a_e \Delta V \sum_{i=1}^{n} N_i = -\Delta W_B$$

Thus, unless the transfer is isothermal, the calculation must be performed by *successive rounds of approximation*, by alternately calculating the transfers of material, heat, material, heat, etc.

4.5. Diffusion of electrolytes

4.5.1. *Chemical potential of electrolytes*

If an electrolyte dissociates to form k anions and a cations with the chemical potentials μ_a and μ_k, the dissociation creates (a + k) ions whose mean chemical potential can be called μ_m. The numbers a and k, respectively, are the *valences* of the anions and the cations.

That portion of the Gibbs energy of the system which pertains to the electrolyte is then:

$$G_{el} = (a + k)\mu_m = k\mu_a + a\mu_k \qquad [4.12]$$

Let us set:

$$\mu_m = \mu_{m0} + RT \, Ln \, \gamma_m \, x_m$$

$$\mu_a = \mu_{a0} + RT \, Ln \, \gamma_a \, x_a \qquad \mu_k = \mu_{k0} + RT \, Ln \, \gamma_k \, x_k$$

By feeding those values back into equation [4.12], μ_{a0} disappears, as does μ_{k0}, if we accept that $\mu_{m0} = (k\mu_{a0} + a\mu_{k0})/(a+k)$.

$$Ln \, \gamma_m = \frac{k \, Ln \, \gamma_a + a \, Ln \, \gamma_k}{a+k}$$

$$Ln \, x_m = \frac{k \, Ln \, x_a + a \, Ln \, x_k}{a+k}$$

If we posit that $a + k = \nu$, we obtain:

$$x_m = \left(x_a^k \, x_k^a \right)^{1/\nu} \qquad \gamma_m = \left(\gamma_a^k \times \gamma_k^a \right)^{1/\nu}$$

and finally, for the activities, which we shall here call $a = \gamma x$:

$$a_m = x_m \gamma_m = \left[(a_a)^k (a_k)^a \right]^{1/\nu} \tag{4.13}$$

Note that x_m is not equal to $(x_a + x_k)$. The molar fraction x_m is a *fictitious* value which enables us to calculate the mean chemical potential of the (a + k) ions.

Whilst this result can easily be generalized to any given number of ionic species, to do so would not hold any apparent interest.

4.5.2. *One type of anion and one type of cation*

The conductive medium (be it a gel or a liquid) in cells and electrolysis baths may be considered to be immobile media where the ions are in motion. However, this motion is less an effect of the variations in concentration than an effect of the electrical field existing between the two electrodes. The result of this is that, as we shall see later on, Fick's equation becomes:

$$N_i = -\frac{D_i}{RT}\left[c_i\frac{d\mu_i}{dz} + c_i z_i F\frac{d\psi}{dz}\right]$$

If ψ is the electrical potential, the electrical field is $-\dfrac{d\psi}{dz}$.

Now consider an electrolyte whose composition is:

$M_a X_k$, made up of M^{k+} and X^{a-} ions

The molecule contains k anions and a cations, whose respective charges are a- and k+ (a for the anion and k for the cation; a and k *are the valences of the ions*).

The charge z_i is equal to $-aF$ or kF where $F = 96.5 \times 10^6$ coulombs.

The molar flux densities are:

$$N_X = -\frac{c_X D_X}{RT}\left[\frac{d\mu_X}{dz} - aF\frac{d\psi}{dz}\right] a > 0 \qquad [4.14]$$

$$N_M = -\frac{c_M D_M}{RT}\left[\frac{d\mu_M}{dz} + kF\frac{d\psi}{dz}\right] k > 0 \qquad [4.15]$$

However, if $M_a X_k$ is the chemical formula of the electrolyte:

$c_M = ac_{M_a X_k}$ and $c_X = kc_{M_a X_k}$

Hence, after adding equations [4.14] and [4.15] together, term by term, we find:

$$\frac{N_X}{D_X} + \frac{N_M}{D_M} = -\frac{c_{M_a X_k}}{RT}\left[k\frac{d\mu_X}{dz} + a\frac{d\mu_M}{dz}\right] = -\frac{c_{M_a X_k}}{RT}\frac{d\mu_{M_a X_k}}{dz}(k+a) = \frac{N_{M_a X_k}}{D_{M_a X_k}}(k+a)$$

Indeed (see section 4.5.1):

$$\mu_{M_aX_k} = \frac{k\mu_X + a\mu_M}{a + k}$$

However:

$$N_X = kN_{M_aX_k} \quad \text{and} \quad N_M = a\,N_{M_aX_k}$$

Hence:

$$\left(\frac{k}{D_X} + \frac{a}{D_M}\right)N_{M_aX_k} = \frac{N_{M_aX_k}}{D_{M_aX_k}}(k + a)$$

The diffusivity of the molecule of electrolyte is therefore given by:

$$D_{M_aX_k} = \frac{D_M D_X (k + a)}{kD_M + aD_X}$$

This expression can be used for transfers of material between two fluids.

4.5.3. *At least three types of ions*

Ions created by the dissociation of an electrolyte diffuse at different velocities. However, the flux of material exchanged between two liquids needs to be neutral. Therefore, an electrical potential is created to correct these disparities. This is known as the "junction potential ψ. "For the ion i, then, the diffusion equation is:

$$N_i RT = -D_i \left[c_i \frac{d\mu_i}{dx} + c_i z_i F \frac{d\psi}{dx} \right] \qquad [4.16]$$

With the exception of the factor c_i, the square bracket represents the electrochemical potential of the ion i.

The z_i values characterize the electrical charge of the ions. The sign of the z_i values is as follows:

$$\text{cations} \quad z_i > 0$$

$$\text{anions} \quad z_i < 0$$

In addition, the flux of material is electrically neutral:

$$\sum_i N_i z_i = 0$$

Term by term, let us add together the diffusion equations after having multiplied each of them by z_i. We find:

$$F\frac{d\psi}{dx} = -\frac{\displaystyle\sum_{i=1}^{n} z_i D_i c_i \frac{d\mu_i}{dx}}{\displaystyle\sum_{i=1}^{n} c_i z_i^2 D_i}$$

This gives us the diffusion equation for each ion:

$$\frac{RT}{D_i} N_i = -c_i \left[\frac{d\mu_i}{dx} - z_i \frac{\displaystyle\sum_{j=1}^{n} z_j D_j c_j \frac{d\mu_j}{dx}}{\displaystyle\sum_{i=1}^{n} c_i z_i^2 D_i} \right]$$

The fact that we have used Fick's law rather than the Maxwell–Stefan equation stems from the hypothesis that, on an ion, the influence of the local electrical field is far greater than that of the particles, independently of their charge.

Remember that the activity coefficients of the ions taken in isolation are given by the method advanced by Helgeson et al. [HEL 81] (see section 3.4.12).

4.5.4. *Neutrality of electrolytic solutions during diffusion*

This neutrality is based on the following two statements:

1) experience shows that any electrolytic solution at equilibrium is electrically neutral everywhere;

2) in the absence of an external electrical field, we have accepted the hypothesis that the electrodiffusive flux density is electrically neutral across any elementary surface.

If we imagine working backwards through time, starting with the solution at its final state of equilibrium (and using the Green–Ostrogradsky theorem), we deduce that throughout the phenomenon of diffusion, the solution has remained electrically neutral at all points. In other words, we always have:

$$\sum_i c_i z_i = 0$$

4.6. Determination of diffusivities

4.6.1. *Measuring the diffusivities*

To measure diffusivity values, the devices described with regard to the complementary equations are used.

4.6.1.1. *Gases*

To determine D^∞ (diffusivity at evanescent concentration – i.e. at infinite dilution), it is sufficient that one of the reservoirs be filled with a dilute mixture of the solute and the other contains only the diluent without solute.

The M.–S equation is written:

$$-c_i \frac{D_{12}}{RT} \frac{\Delta \mu_i}{\Delta z} = (N_i x_j - N_j x_i) \ (j \text{ or } i = 1 \text{ or } 2 \text{ and } i \neq j)$$

By positively orienting the fluxes from gas A toward gas B:

$$-\Delta \mu_i = \mu_{iA} - \mu_{iB} = \frac{\Delta z}{c_i} \frac{RT}{D_{12}} (N_i x_j - N_j x_i)$$

Because the mixture is gaseous, the complementary equation is:

$$\sum_i N_i = 0 \quad \text{(see section 4.2.3.2)}$$

During a (very short) period of time $\Delta\tau^{(1)}$, analysis of the contents of the reservoirs shows that the quantity Δn_i kilomoles of gas i has shifted from one reservoir to the other, and we have:

$$N_i = \frac{\Delta n_i}{A_T \Delta\tau^{(1)}} \quad (\text{kmol.m}^{-2}.\text{s}^{-1}) \qquad [4.17]$$

A_T : section of the tube: m^2;

$\Delta\tau^{(1)}$: time interval: s;

Δz : length of mixture in the tube: m.

Hence:

$$c_i(\mu_{iA} - \mu_{iB}) = \frac{RT}{D_{12}\Delta\tau^{(1)}}\left(\frac{\Delta z}{A_T}\right)(\Delta n_i x_j - \Delta n_j x_i) \quad (\text{i or } j=1 \text{ or } 2 \text{ and } i \neq j)$$

The ratio $\Delta z/A_T$ is determined with a couple of gases for which we know D_{12} and for which we work with the same time period $\Delta\tau^{(1)}$ as for the couple under study. The value A_T in equation [4.17] shall be the real value.

Furthermore, by knowing Δn_i, we are able to find the compositions and the chemical potentials which are evaluated using the arithmetic means of the compositions at the start and end of the time interval $\Delta\tau^{(1)}$.

The fact that the analyses require readings in both reservoirs means that, for each $\Delta\tau^{(k)}$, the experiment is restarted from the beginning.

Thus, the experiment with index (k) would last for:

$$\tau^{(k)} = \sum_{j=1}^{k} \Delta\tau^{(j)}$$

The readings for analysis are only taken at time $\tau^{(k)}$.

4.6.1.2. *Liquids*

The procedure is the same as for gases.

The reservoirs are kept homogeneous with magnetic stirrers.

For liquid mixtures, the complementary equation is:

$$\sum_i N_i M_i = 0 \quad \text{(see section 4.2.3.1)}$$

where the M_i are the molar masses.

4.6.2. *Predictive calculation of the diffusivities in liquids (non-electrolytes):*

Values are given in Loncin [LON 85], and a predictive calculation can be performed using the method given by Wilke and Chang [WIL 55] who, for diffusivity, give the following expression:

$$D_\infty = \frac{7.4\times10^{-12}\,T\,(XM)^{0.5}}{\mu\,V^{0.6}}$$

 – μ: viscosity of the solution: cp

 – T: absolute temperature: K

 – X: association parameter of the solvent

 – M: molar mass of the solvent: kg.kmol^{-1}

 – V: molar volume of the solute: L.kmol^{-1} (see Table 4.1)

In [WIL 55], readers will find a method for calculating the molar volumes of organic species.

D_∞: diffusivity at infinite dilution: m^2.s^{-1}

For associated solvents, the parameters X are:

Water   2.6

Methyl alcohol   1.9

Ethyl alcohol   1.5

EXAMPLE 4.1.–

Diffusivity of oxygen in water.

$$T = 293K \qquad V = 25.6 \text{ L.kmol}^{-1} \qquad M = 18 \text{kg.kmol}^{-1}$$

$$\mu = 1 \text{ cp} \qquad\qquad\qquad X = 2.6$$

$$D = \frac{7.4. \times 0^{-12} \times 293 \times (2.6 \times 18)^{0.5}}{1 \times 25.6^{0.6}}$$

$$D = 2.11 \times 10^{-9} \text{m}^2.\text{s}^{-1}$$

Air	29.9	H_2S	32.9
Br_2	53.2	I_2	71.5
Cl_2	48.4	N_2	31.2
CO	30.7	NH_3	25.8
CO_2	34	NO	23.6
COS	51.5	N_2O	36.4
H_2	14.3	O_2	25.6
H_2O	18.9	SO_2	44.8

Table 4.1. *Molecular volumes of liquids (L.kmol.$^{-1}$; [TRE 80])*

NOTE.– 1) Scheibel [SCH 64] proposed a correlation not involving an association parameter.

2) If we know the diffusivities at infinite dilution, we propose to write, as Vignes [VIG 66] does:

$$D_{ij} = (D_{ij}^{\infty})^{q_{ji}} \times (D_{ji}^{\infty})^{q_{ij}} = D_{ji}$$

where:

$$q_{ji} = \frac{x_j}{x_i + x_j} \quad \text{and} \quad q_{ij} = \frac{x_i}{x_i + x_j}$$

D_{ij}^{∞} : diffusivity of i infinitely dilute in j: $m^2.s^{-1}$;

D_{ji}^{∞} : diffusivity of j infinitely dilute in i: $m^2.s^{-1}$.

If we know only D_{ij} at two arbitrary concentrations, it is easy to deduce D_{ij} and D_{ji} corresponding to the infinite dilutions by turning to logarithms.

4.6.3. Stokes–Einstein equation (large molecules)

This expression of the diffusivity is imposed when the solute is a large molecule of spheroidal form (in solution).

The volume of such a molecule is estimated thus:

$$V = \frac{M}{N_A \rho_L}$$

M: molar mass: $kg.kmol^{-1}$;

N_A: Avogadro's number: 6.023×10^{26} molecule.$kmol^{-1}$;

ρ_L: density of the solute in the pure state and liquid (or solid): kg m^{-3}.

The molecular radius is then:

$$r = \left(\frac{3V}{4\pi}\right)^{1/3}$$

The Stokes–Einstein law is then written as follows, for high dilution:

$$D_\infty = \frac{k_B T}{6\pi r \eta}$$

k_B: Boltzmann's constant: 1.38048×10^{-23} J.K^{-1};

η: viscosity of the solvent: Pa.s;

T: absolute temperature: K.

With regard specifically to proteins, Young et al. [YOU 80] propose an expression for diffusivity:

$$D = 7.51 \times 10^{-15} \left(\frac{T}{\eta V^{1/3}} \right)$$

NOTE.– [PER 36] studied the diffusivity of ellipsoidal molecules.

4.6.4. Reciprocal diffusivity in gases

The diffusivity of gas i in gas j is equal to the diffusivity of gas j in gas i – hence the term "reciprocal diffusivity".

The expression given by Gilliland [GIL 34] for reciprocal diffusivity, confirmed by the measurements taken Fuller et al. [FUL 66] is indeed symmetrical in terms of i and j.

$$D_{ij} = \frac{0.043 \times 10^{-5} T^{1.5} \sqrt{1/M_i + 1/M_j}}{P \left(V_i^{1/3} + V_j^{1/3} \right)^2}$$

T: absolute temperature: K;

M_i and M_j: molar masses: kg.kmol^{-1};

V_i and V_j: molar volumes: L.kmol^{-1};

P: pressure: bar abs.

EXAMPLE 4.2.–

Reciprocal diffusivity of air and water vapor.

$$V_1 = 20.1 \text{ L.kmol}^{-1} \qquad P = 1 \text{ bar abs} \qquad M_1 = 29 \text{ kg.kmol}^{-1}$$

$$V_2 = 12.7 \text{ L.kmol}^{-1} \qquad T = 293 \text{ K} \qquad M_2 = 18 \text{ kg.kmol}^{-1}$$

$$D = \frac{0.043 \times 10^{-5} \times T^{1.5} \sqrt{1/29 + 1/18}}{1\left(20.1^{1/3} + 12.7^{1/3}\right)^2} = 2.54 \times 10^{-5} \text{ m}^2 \text{.s}^{-1}$$

NOTE.– Fuller *et al.* [FUL 66] proposed to replace the factor $0.043 \times 10^{-5}.T^{1.5}$ with the factor $10^{-7} T^{1.75}$. They give the increments of structural volumes for complicated molecules.

H_2	7.07	CO	18.9
D_2	6.70	CO_2	26.9
He	2.88	$N_2 O$	35.9
N_2	17.9	NH_3	14.9
O_2	16.6	H_2O	12.7
Air	20.1	CCl_2F_2	114.8
Ar	16.1	SF_6	69.7
Kr	22.8	Cl_2	37.7
Xe	37.9	Br_2	67.2
		SO_2	41.1

Table 4.2. *Molecular volumes of gases (L.kmol.$^{-1}$; [FUL 96])*

4.6.5. *Equivalent conductivity and molar conductivity of ions*

The conductance G of a liquid (or a solid) is the inverse of its resistance and its measured value in siemens:

$$G = \frac{1}{R} \qquad (1S = 1\Omega^{-1})$$

$$[G] = \frac{A}{V} = \frac{\text{coulomb}}{\text{s x volt}} = C.s^{-1}.V^{-1}$$

The conductivity σ is (see Appendix 9):

$$\sigma = \frac{GL}{\Sigma} \qquad (S.m^{-1})$$

L: length of the conductor: m;

Σ: section of the conductor: m².

We can therefore obtain the equation of the conductivity dimensions:

$$[\sigma] = C.s^{-1}.m^{-1}.V^{-1} = A.V^{-1}.m^{-1} \text{ (A for amperes)}$$

The conductivity per equivalent present in the liquid at the concentration $c_{eq} keq.m^{-3}$ is:

$$[\lambda_{eq}] = \frac{\sigma}{c_{eq}} = \frac{m^2 A}{keq .V} = \frac{m^2 S}{keq} \text{ (keq for kiloequivalent)}$$

By definition, the conductivity λ_{eq} is positive.

The ionic conductivity of the ion i is proportional to its valence z_i.

$$\lambda_{ion\ i} = |z_i| \lambda_{eqi}$$

The molar conductivity of an electrolyte is proportional to the number of equivalents present in a mole:

$$\Lambda_{mol} = \Sigma v_i |z_i| \lambda_{eqi} = \Sigma v_i \lambda_{ion\ i} \text{ (m}^2.\text{kmol}^{-1}.\text{S.)}$$

v_i : stoichiometric coefficient of the ion i: ions.mol^{-1};

$|z_i|$: absolute value of the valence of the ion i: equ.ion^{-1}.

The conductivity of a solution containing c kmoles of electrolyte per m³ is:

$$\kappa = c\Lambda \text{ where } [\kappa] = S.m^{-1}$$

This result is consistent with the general definition of the conductivity σ which is measured in Ω.m.

The equivalent conductivity of the solution is:

$$\Lambda_{eq} = \frac{\kappa}{c\sum_i v_i |z_i|} = \frac{\sum_i v_i |z_i| \lambda_{eqi}}{\sum_i v_i |z_i|} \qquad (m^2.keq^{-1}.S)$$

The diffusivity of the ion is:

$$D_{ion} = \frac{RT\lambda_{ion\ i}}{z_i F^2} \qquad\qquad [4.18]$$

4.6.6. Conductance, mobility and diffusivity

The mobility of the ion i is defined by:

$$u_i = \frac{z_i \lambda_{eqi}}{F} \qquad (m.s^{-1}(V.m^{-1}) = m^2.s^{-1}.V^{-1})$$

F: Faraday

$$F = 96.487 \times 10^6 \text{ Coulomb.kiloequivalent}^{-1} = 96.487 \times 10^6 \text{ C.keq}^{-1}$$

The mobility of the anions is negative because anions move in the opposite direction to the electrical field.

The mobilities of different ions of the same valence do not differ greatly, except for those of H^+ and OH^-, whose mobilities are six times greater than the respective mean values of the cations and the anions.

Li^+ and Na^+ have low mobility because they are highly hydrated.

Let us verify the equation with the dimensions of the mobility u_i:

$$\left[\frac{\lambda_i}{F}\right] = \frac{m^2 G \text{ keq}}{\text{keq C}} = \frac{m^2 \text{ C}}{\text{C s V}} = \frac{m^2}{sV} = [u_i]$$

The diffusivity is defined by:

$$D_i = \frac{RTu_i}{z_i F} \qquad\qquad\qquad [4.19]$$

where:

$$[RT] = \frac{J}{\text{keq}} = \frac{CV}{\text{keq}}$$

Hence:

$$[D_i] = \frac{C.V.m^2.\text{keq}}{\text{keq.s.V.C}} = \frac{m^2}{s}$$

EXAMPLE 4.3.–

Diffusivity of calcium chloride at 18°C (k = 2 and a= 1)

$$u_{Ca^{++}} = 1.08 \times 10^{-7} m^2.s^{-1}.V^{-1} \qquad\qquad u_{Cl^-} = 0.68 \times 10^{-7} m^2.s.V^{-1}$$

In Table II.9 of Vol. I of his book, [MIL 69] claims to give the mobility u_i of the ions. In reality, he gives the ratio u_i/z_i.

$$\frac{1}{2}\lambda_{eqCa^{++}} = 5.21 m^2.\text{S. keq}^{-1}.m^2 \quad \lambda_{eqCl^-} = 6.56 m^2.\text{S. keq}^{-1}.m^2$$

In light of equation [4.18]:

$$D_{Ca^{++}} = \frac{8314 \times 291.15 \times 5.21}{(96.487 \times 10^6)^2} \qquad\qquad D_{Cl^-} \frac{8314 \times 291.15 \times 6.56}{(96.487 \times 10^6)^2}$$

$$D_{Ca^{++}} = 1.355 \times 10^{-9} m^2.s^{-1} \qquad\qquad C_{Cl^-} = 1.705 \times 10^{-9} m^2.s^{-1}$$

$$D_{CaCl^2} = \frac{1.355 \times 1.705 \times 10^{-9} \times 3}{2 \times 1.355 + 1.705}$$

$$D_{CaCl^2} = 1.57 \times 10^{-9} m^2.s^{-1}$$

4.6.7. Transport number, transference

The transport number of an ion i within an immobile electrolytic solution traversed by an electric current is the fraction of that current transported by the ion in question.

$$T_i = \frac{z_i u_i c_i}{\sum\limits_{j}^{n} z_j u_j c_j}$$

As the product $z_i u_i$ must always be positive, the transport number too is always positive. In the literature, we often find different definitions for the transport numbers. For example:

$$\tau_i = \frac{u_i z_i}{\sum\limits_{i} u_i z_i}$$

The transference τ_i of an ion i is the number of kilomoles of i transported by the fraction of current T_i.

$$\tau_i = \frac{T_i}{z_i}$$

The transference of an anion is, of course, negative.

4.7. Concepts concerning batteries and electrolytic cells

4.7.1. Galvani potential

This potential Φ is a potential within an α phase and represents the electrical work needed to bring a charge of 1 Coulomb from infinity in a vacuum (or $\Phi = 0$) into the α phase.

From infinity to the vicinity of the surface of the α phase, the potential needing to be created is ψ. This is the Volta potential. Next the charge needs to cross the surface which usually contains dipoles creating a surface potential χ, so the Galvani potential of the charge in the α phase is:

$$\Phi_{Galvani} = \psi_{Volta} + \chi_{surface}$$

4.7.2. Electrochemical potential

The electrochemical potential of an ion within the α phase, then, is:

$$\tilde{\mu_i} = \mu i + z_i F\Phi = \mu_{oi}(T, P) + RTLna_i + z_i F\Phi :$$

– Φ: Galvani potential: volts;

– F: Faraday's constant: 96.487×10^6 coulombs per kilovalence;

– z_i: "valence" of the ion (number of kilovalences per kiloion). This number is negative with anions;

– a_i: activity of the ion: $a_i = \gamma_i x_i$.

4.7.3. The battery using a hydrogen electrode

A battery is generally composed of two electrodes in contact with one another through two electrolytic solutions. By convention, the Galvani potential of the hydrogen electrode is taken as equal to zero, on the understanding that this electrode is at equilibrium with gaseous hydrogen at the pressure of 1 atm.abs.

Around the other electrode, an ion (or indeed a molecule) is likely to be reduced by the addition of electrons via *the external circuit*, which is a simple metal wire linking the two electrodes throughout their use.

A battery exploits a reduction/oxidation (redox) reaction. The electrode where oxidation takes place is the anode, and the cathode is the site of reduction. (Cathode: Reduction – Anode: Oxidation gives us the mnemonic "CRAO"). The same is true of an electrolysis cell.

1) The following two reactions occur in the cell:

oxidation of hydrogen (anode, −)

$$\frac{1}{2}H_2 \rightarrow H^+ + e$$

reduction of the species X (cathode, +)

$$X + e \rightarrow X^-$$

Thus, in total:

$$\frac{1}{2}H_2 + X \rightarrow H^+ + X^-$$

2) In an electrolysis cell, the direction of the arrows is reversed.

reduction of the cation H^+ (cathode, −)

$$H^+ + e \rightarrow \frac{1}{2}H_2$$

oxidation of the anion X^- (anode, +)

$$X^- \rightarrow X + e$$

Thus, in total:

$$H^+ + X^- \rightarrow \frac{1}{2}H_2 + X$$

For each of these two reactions, there is a corresponding variation of the Galvani potential for the reactant species. Let these variations be represented as $\Delta\Phi_{H2}$ and $\Delta\Phi_X$. By convention, it is agreed that $\Delta\Phi_{H2}$ is zero, and therefore we write the electromotive force of the cell as follows.

$$E = \Delta\Phi_X - \Delta\Phi_{H2} = \Delta\Phi_X$$

Outside of the cell, it is electrons which circulate, and inside, it is ions. As we know, the direction of the current in the external circuit is the opposite to the direction of movement of the electrons. If $\Delta\Phi_X$ is positive, the half-cell X receives electrons and gives out negative ions. This half-cell contains the cell's + pole: the anode. The cathode has the sign −.

A positive value of $\Delta\Phi_X$ means that the species in question is easily reduced, and is therefore oxidizing. Conversely, a negative variation $\Delta\Phi_X$ corresponds to a highly-reductive species (readily donating electrons, as the alkali metals do).

Note that, by generalizing, we can state that the overall reaction of any cell is the sum of a reduction reaction and an oxidation reaction, and therefore that:

$$E = \Delta\Phi_{red} - \Delta\Phi_{ox}$$

The half-cell corresponding to the reduction (acceptance of electrons) *contains the terminal marked + and corresponds to the higher potential* $\Delta\Phi$. This is the anode. The cathode corresponds to oxidation, and bears the sign −.

In an electrolysis cell, the direction of the current is inverted and the half-cells keep their names. The anode has the sign + and the cathode the sign −. The metal ion is then discharged at the cathode − hence its name: cathion (or cation).

4.7.4. Nernst's equation

Let z represent the number of electrons involved in each of the two reactions. Indeed, in order for the global reaction to be balanced, the two redox reactions must involve the same number of electrons. The product zF, then, is the electrical charge exchanged *in the cell* between the two electrodes, the direction of the current (rather than the direction of movement of the electrons) flows in the opposite direction to the external circuit − i.e. from the oxidation electrode toward the reduction electrode.

The electrical energy corresponding to this shift of charges is:

$$W = EzF > 0$$

However, the spontaneous, global reaction corresponds to a decrease in the chemical Gibbs energy. Thus:

$$\Delta G_{chem} = -EzF < 0$$

The reaction expressing the operation of a cell is, generally:

$$aA + bB \qquad cC + dD \rightarrow$$

The variation of the chemical Gibbs energy ΔG_{chem} corresponding to this reaction is:

$$\Delta G_{chem} = \sum_i v_i \mu_{0i} + \sum_i v_i RTLna_i = \Delta G_0 + \sum_i v_i RTLna_i$$

The stoichiometric coefficients v_i, here, are negative for the reagents and positive for the products.

Finally, Nernst's equation is written as follows, by equaling ΔG_{chem} and the work of the electrons.

$$\Delta G_0 + RTLn \frac{a_C^c a_D^d}{a_A^a a_B^b} = - vFE \quad (a, b, c \text{ and } d \text{ are positive})$$

v is the number of electrons involved in the reaction.

Note that:

$$2.303 \times \frac{RT}{F} = \frac{8315 \times 298.15 \times 2.303}{96485} = 0.059$$

At equilibrium:

$$E = \frac{0.059}{v} \log_{10} K$$

NOTE.– The electrochemical potential of the reaction remains invariable:

$$\Delta G_{chem} + vF\Delta\Phi = 0 \quad \text{where} \quad \Delta\Phi = E$$

The decrease in the chemical Gibbs energy corresponds to an increase in the Galvani potential of the reaction.

Calculation of the Equilibrium Between Two Fluids

5.1. General

5.1.1. *General equilibrium between two phases*

Consider an isolated system which is the union of the two phases. We know that:

$$dU = TdS - PdV + \sum_{i=1}^{c} \mu_i dn_i$$

Hence:

$$dS = \frac{1}{T} dU + \frac{P}{T} dV - \sum_{i=1}^{c} \frac{\mu_i}{T} dn_i$$

For the system that is the union of the two phases A and B:

$$U = U_A + U_B = \text{const.}$$

$$V = V_A + V_B = \text{const.}$$

$$n = n_A + n_B = \text{const.}$$

Thus:

$$dU_B = -dU_A$$

$$dV_B = -dV_A$$

$$dn_B = -dn_A$$

At equilibrium, the entropy of the whole system is maximal:

$$dS = dS_A + dS_B = 0$$

This means that:

$$dS = \left[\frac{1}{T_A}dU_A + \frac{P_A}{T_A}dV_A - \frac{\mu_A}{T_A}dn_A\right] + \left[\frac{dU_B}{T_B} + \frac{P_B}{T_B}dV_B - \frac{\mu_B}{T_B}dn_B\right] = 0$$

$$dS = \left(\frac{1}{T_A} - \frac{1}{T_B}\right)dU_A + \left(\frac{P_A}{T_A} - \frac{P_B}{T_B}\right)dV_A - \left(\frac{U_A}{T_A} - \frac{U_B}{T_B}\right)dn_A = 0$$

This equation must be satisfied for arbitrary values of dU_A, dV_A and dn_A. Therefore:

$$T_A = T_B$$

$$P_A = P_B$$

$$\mu_A = \mu_B$$

Often, the chemical potential of a given species of index i can be written using the activity a_i of that species:

$$\mu_i = \mu_{oi}(T,P) + RTLna_i$$

For an equilibrium between two fluids A and B, this gives us:

$$a_{Ai} = a_{Bi}$$

We can also write, using the fugacity f_i:

$$\mu_i = \mu_{oi}(T) + RTLnf_i$$

and the equilibrium is then written:

$$f_{Ai} = f_{Bi}$$

5.1.2. *Phase rule*

Consider the system containing c components distributed between φ phases. This system is entirely defined by the $(c-1)1$ molar fractions giving the composition of each phase – thus we have $\varphi(c-1)$ variables of state, in addition to which we must consider the pressure and temperature. These variables are linked by $c(\varphi-1)$ equilibrium relations (equality of the chemical potentials for each component). The number of parameters left free for the definition of the system is the variance v of that system:

$$v = \varphi(c-1) + 2 - c(\varphi-1) = c + 2 - \varphi$$

5.1.3. *Ratio of molar fractions at equilibrium*

Regarding the component of index i, the ratio at equilibrium is defined by:

$$E_i = \frac{y_i}{x_i}$$

For the operations of distillation, absorption and stripping, y_i is the molar fraction of i in the gaseous phase and x_i is the molar fraction in the liquid phase.

In the case of liquid–liquid extraction, y_i is the molar fraction of i in the solvent phase (we also speak of the extract). The fraction x_i pertains to the raffinate or phase or residue, depending on whether the noble product is in the raffinate or in the extract.

It is common to write:

$$\mu_i = \mu_i(T,1) + RTLn\hat{f}_i$$

where:

$$\hat{f}_i = x_i \gamma_i \varphi_i P = x_i \gamma_i f_i = x_i \hat{\varphi}_i P$$

– $\hat{f}_i$: fugacity in the mixture: Pa

– φ_i : fugacity coefficient of component i in the pure state

– P : total pressure: Pa

– f_i : fugacity of component i in the pure state: Pa

– If $\hat{\varphi}_i$ is the fugacity coefficient of component i in the mixture, we deduce:

$$\gamma_i = \frac{\hat{\varphi}_i}{\varphi_i}$$

Brunner [BRU 94] gives the descriptions of numerous equilibria between fluids and, in particular, supercritical equilibria.

5.2. Representation of liquid–vapor equilibria

5.2.1. *Distinction between liquid phase and gaseous phase*

Gases always occupy the *whole* of the volume available to them, regardless of the possible presence of other gases. Therefore, we can only speak of the volume as an independent variable. The total pressure of the mixture is a resultant of the available volume, the quantity of gas present and the temperature.

A liquid, on the other hand, only *partially* occupies the volume available to it, and its pressure does not depend on that volume because it is equal to its vapor pressure. The volume not occupied by the liquid is occupied by its saturating vapor. The only independent variable here is the temperature – that is to say, the volume of the liquid depends only on the amount of fluid present and on the temperature, rather than on the total volume available (if, of course, we ignore the mass of the vapor in comparison to that of the liquid). Hence, it is natural that, when dealing with a liquid mixture, we should use the concept of partial volume, whereas for a gas, we use the concept of partial pressure (which is identical to fugacity in the case of perfect gases).

5.2.2. *Behavior of two phases of the same composition*

Consider the liquid- and vapor phases of a binary mixture, and let 1 represent the less volatile component and 2 the more volatile.

Let z_1 be the molar fraction of component 1 (the less volatile).

For T = const., the saturating vapor pressure of component 2 is greater than that of component 1.

$$\pi_1(T) < \pi_2(T)$$

For P = const., the boiling point τ_1 of 1 is greater than that of 2:

$$\tau_1(P) > \tau_2(P)$$

From this, we can deduce the shape of the pressure- and temperature bundles.

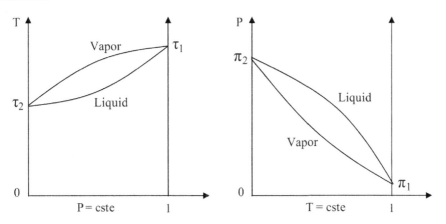

Figure 5.1. *Pressure and temperature bundles*

For a given composition z for the liquid and the vapor, one of the bundles gives the boiling- and dew-point temperatures, and the other the boiling- and dew-point pressures. Of course, the phases thus defined *are not at mutual equilibrium*. The compositions of the two phases at equilibrium are at the intersection of the bundles with a horizontal.

Note that, for T = const., the boiling pressure P_L of the liquid is greater than the dew pressure P_V of the gaseous mixture. We can deduce the shape of the isotherms characterizing a mixture of fixed composition z in comparison to the isotherms of a pure substance.

If the solution is not ideal, we may see the formation of a positive azeotrope for which:

– the temperature is less than T_1 and T_2;

– the pressure is greater than π_1 and π_2;

– A negative azeotrope may also appear and, for that azeotrope:

  - the temperature is greater than T_1 and T_2,

  - the pressure is less than π_1 and π_2.

If the pressure P is greater than the critical pressure of one of the two components, the bundle does not reach the ordinate axis corresponding to that component.

In Figure 5.2, the pressure P is greater than the critical pressure P_{C1} of component 1, and less than the critical pressure P_{C2} of component 2:

$$P_{C1} < P < P_{C2}$$

The supercritical fluid, here, is fluid 1.

In Figure 5.3, the constant temperature T is greater than the critical temperature T_{C2} of component 2 in the pure state, but remains less than that of component 1. This means that fluid 2 is in the supercritical state.

$$T_{C2} < T < T_{C1}$$

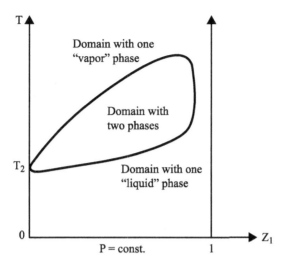

Figure 5.2. *Mixture supercritical in relation to component 1*

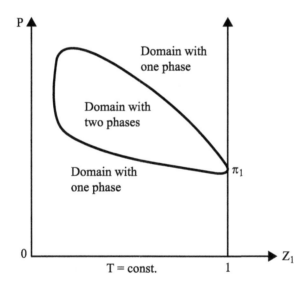

Figure 5.3. *Mixture supercritical in relation to component 2*

Regarding the liquid–vapor equilibrium of a mixture, we need to apply the laws of thermodynamics and write the equality of the chemical potentials of each of the components in the two phases. If we begin by taking the pressure and the liquid fraction L, we see that the liquid and the gas have different compositions at the common temperature sought, but that the chemical potentials are equal:

$$\mu_i^V = \mu_i^L = \mu_i$$

$$G^V = \sum y_i \mu_i^V = \sum y_i \mu_i \neq \sum x_i \mu_i = \sum x_i \mu_i^L = G^L$$

Thus:

$$G^V \neq G^L$$

If we now assimilate the mixture to a pure substance, we can calculate the boiling point T_{boil} of that *pure substance* at the imposed pressure. Thus, at T_{boil} and $\pi(T_{boil}) = P$, we have:

$$G^V = G^L$$

Unfortunately, this way of working is wrong because, except for an azeotrope, a liquid mixture at equilibrium with a gaseous mixture has a composition different to that of the gas.

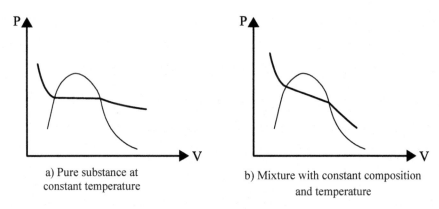

a) Pure substance at
constant temperature

b) Mixture with constant composition
and temperature

Figure 5.4. *Isotherms* $P = f(V, T)$

A third approach is, still at a given pressure, to look for:

– the dew point of the mixture by making $L = 0$;

– the boiling point of the mixture by making $L = 1$.

We see that, for a given pressure:

$$T_{dew} > T_{boil} \tag{5.1}$$

For a given temperature:

$$P_{dew} < P_{boil} \tag{5.2}$$

Figure 5.4(b) corresponds to the inequality [5.2], but the two phases which have the same composition and same temperature *are not at mutual equilibrium*, because they are not at the same pressure.

5.2.3. *Finding the nature of certain points of equilibrium*

It may happen that we need to precisely examine a phenomenon of retrograde condensation.

By way of example, consider the mixture represented in Figure 5.5. We can see that the cricondentherm and cricondenbar are both on the dew curve. This means that at the same temperature, the mixture has two dew points (points A and B) and at the same pressure, it also has two dew points (points D and B).

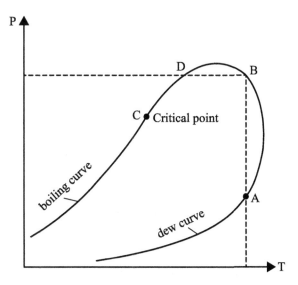

Figure 5.5. *Retrograde condensation*

It is possible to identify each of the points A, B or D on the basis of the sign of certain derivatives, as shown by Table 5.1:

Point	P	T	N_V	$(\partial P/\partial T)_{N_V}$	$(\partial N_V/\partial T)_P$
A	P_A	T_A	1	> 0	> 0
B	P_B	$T_B = T_A$	1	< 0	> 0
D	$P_D = P_B$	T_D	1	> 0	< 0

Table 5.1. *Identification of points by derivatives*

5.2.4. *Liquid–vapor equilibrium at constant composition*

If, for a given composition, we represent the liquid–vapor curve with T on the abscissa axis and P on the ordinate axis, we obtain the curve $f(P,T) = 0$.

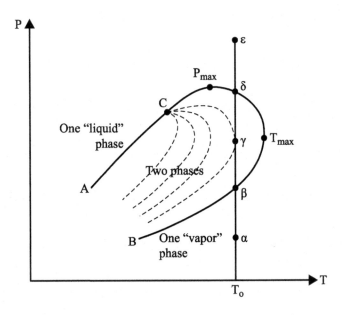

Figure 5.6. *Pressures and temperatures for a constant composition*

The dotted curves each define a constant "quality" (the quality is the vaporized fraction of the mixture). Thus, the domain falling within the curve $f(P,T) = 0$ corresponds to two-phase mixtures. Outside of that curve, however, only one phase exists.

The point C at which all the quality curves end is the critical point corresponding to the composition of the mixture. At that point, the liquid and the vapor phase are indiscernible. Conventionally, on the arc CB, we are dealing with a vapor and, on the arc AC, with a liquid. This disposition enables us to describe the phenomena of retrograde vaporization and condensation.

Indeed, consider a "vapor" phase α and increase its pressure P whilst maintaining $T = const. = T_0$. Up until point β, nothing happens. Then, between β and γ, partial liquefaction occurs. However, between γ and δ, the liquefied fraction disappears. This vaporization, which is caused by an increase in pressure, is known as retrograde vaporization. Had we followed the path in the opposite direction (from ε to α), then between δ and γ we would have seen liquefaction because of the drop in pressure – i.e. retrograde liquefaction. Between γ and β, vaporization would have occurred normally because of expansion. A similar phenomenon would have been observed if we had operated at $P = const. = P_0$.

Depending on the nature of the components of the mixture, the critical point may be found:

– between P_{max} and T_{max};

– outside of the interval $[P_{max}, T_{max}]$:

   - on the side of T_{max},

   - or on the side of P_{max}.

5.2.5. *Ideal mixtures*

From the uniformity of the molecular interactions, it stems that the extensive molar values of state of ideal mixtures are of the form:

$$m^{id} = \sum_i x_i m_i$$

The m_i are the values of state of the pure liquid components at T and P of the mixture. Specifically, these are:

– the volumes $v_i(T)$;

– the internal energies $u_i(T)$;

– the enthalpies $h_i(T)$.

5.3. Calculation of the equilibria between two fluids

5.3.1. *Partial material balances and overall balance*

We shall look at the example of the liquid–vapor equilibrium, but the calculations below *remain valid for liquid–liquid equilibria*, if we replace L with R (raffinate or residue) and V with E (extract).

If we perform the equilibrium calculation for 1 kilomole of mixture, the balances are written:

$$z_i = Lx_i + Vy_i$$

z_i : molar fraction of the component i in the overall mixture.

In all the discussions below, obviously we suppose that the system's overall composition (the z_i values) is known.

There are n partial material balance equations and a global balance equation:

$$L + V = 1$$

Hereafter, we shall eliminate L from the start by replacing it with $(1 - V)$.

5.3.2. *Equilibrium equations*

The ratios at equilibrium express the equality of the fugacities f_i^V and f_i^L (see section 5.1.1).

$$E_i = \frac{y_i}{x_i} = \frac{\hat{\varphi}_i^L}{\hat{\varphi}_i^V} \qquad \text{or indeed} \qquad E_i = \frac{y_i}{x_i} = \frac{\gamma_i \pi_i}{\varphi_i^V P}$$

$\hat{\phi}_i$: fugacity coefficient of the species i in a mixture

γ_i : activity coefficient

π_i : vapor pressure: Pa

We can see that the ratios at equilibrium are defined by the $2n + 2$ variables which are P, T, x_i and y_i.

If the temperature is much greater than the critical temperature T_{cj}, we can use Henry's constant $H_j(T)$.

$$E_j^{(0)} = \frac{H_j(T)}{P}$$

For a liquid–liquid equilibrium, we write (see section 5.1.1)

$$a_i^R = a_i^E \quad \text{which means that} \quad E_i = \frac{y_i}{x_i} = \frac{\gamma_i^{R'}}{\gamma_i^E}$$

5.3.3. Normalization equation

We may choose any one of the following three equations at will:

$$\sum_1^n x_i = 1$$

$$\sum_1^n y_i = 1$$

$$\sum_1^n (x_i - y_i) = 0$$

If one of the three equations is satisfied, the other two automatically will be, if we take account of the partial material balances and the global balance.

The third equation has the advantage of making the two phases play a symmetrical role.

5.3.4. *The system resolving the (2n + 1) equations*

The partial material balance of the component i is written:

$$z_i = Lx_i + Vy_i = (1-V)x_i + VE_ix_i$$

Therefore, we have 2n equations:

$$x_i = \frac{z_i}{1-V+E_iV} \quad \text{and} \quad y_i = \frac{E_iz_i}{1-V+E_iV}$$

These values of y_i replace the equilibrium equations.

Yet:

$$\sum_1^n (x_i - y_i) = 0$$

From this, we deduce the "N" equation, which replaces the normalization equation.

$$N(T,P,V) = \sum_1^n \frac{(1-E_i)z_i}{1-V+E_iV} = 0$$

Therefore, we do indeed have 2n+1 equations.

5.3.5. *Available data and unknowns to be calculated*

The system is entirely determined if we know:

– the n molar fractions x_i in the liquid;

– the n molar fractions y_i in the vapor phase;

– the 3 values of state, P, T and V.

Because we have only (2n + 1) equations, it is necessary to employ two additional relations. The following four problems are the most common:

– we take P and T and search for the value of V;

– we take V and P and search for the value of T;

– we take P and H (the enthalpy) and search for two unknowns V and T but, in this particular case, we have the additional functional:

$$H = (V, x_i, y_i, T, P)$$

– we take P and S (the entropy) and search for two unknowns V and T, using the functional:

$$- S = S(V, x_i, y_i, T, P).$$

The calculation procedure for the last two cases is identical. Therefore, we shall merely detail the solution of the first three problems.

5.3.6. *We take P and T and look for V*

The procedure is as follows:

1) An initial estimation of the E_i values can be obtained by accepting the laws of ideality – i.e.:

$$E_i^{(0)} = \frac{\pi_i(T)}{P} \qquad (\pi_i \text{ is the vapor pressure of the substance i})$$

2) The resolving system then gives $V^{(0)}$, $x_i^{(0)}$ and $y_i^{(0)}$.

3) We calculate $E_i^{(1)}$ as a function of $V^{(0)}$, $x_i^{(0)}$ and $y_i^{(0)}$.

4) The resolving system gives $V^{(1)}$, $x_i^{(1)}$, $y_i^{(1)}$.

etc.

We establish the ratio:

$$r^{(n)} = \frac{V^{(n-1)} - V^{(n)}}{V^{(n)}}$$

The ratio $r^{(n)}$ must be less than 10^{-5} in order for $V^{(n)}$ to be acceptable.

The value found for V must be between 0 and 1. If we find V > 1, we are dealing with a superheated gas, and we accept that V = 1. If V < 0, it is a supercooled liquid, and we write V = 0.

At each step, before calculating the E_i values, it is necessary to *normalize* the x_i and y_i by writing:

$$x_i = \frac{x_i}{\sum_i x_i} \qquad \text{and} \qquad y_i = \frac{y_i}{\sum_i y_i}$$

Indeed, as long as the solution is not achieved, there is no reason for the sums of the x_i and y_i to be equal to 1.

NOTE.– We could use the digital method shown in Appendix 5 to solve the "N" equation, but it is also noteworthy that, by reduction to the same denominator, the solutions to that equation are the roots of a polynomial whose degree is equal to the number of components of the mixture. The resolution of the third- and fourth-degree equations is given in Appendix 2. We need to choose the root that lies between zero and 1.

5.3.7. *We take V and P and look for T*

1) If we know the boiling points τ_i of the different components at pressure P, we calculate an initial estimation of T:

$$T^{(0)} = \sum_i z_i \tau_i$$

2) By applying the procedure 5.3.6, we obtain $V^{(0)}$, $x_i^{(0)}$, $y_i^{(0)}$

3) More generally, what we need to do is eliminate F(T) with:

$$F(T) = V(T) - V$$

Appendix 5 shows the tangential method (which is helpful for this problem). V is an increasing monotonic function of T.

Note that if we impose $V = 1$, the calculation gives the dew point of the mixture. Similarly, with $V = 0$, we find the boiling point.

5.3.8. *Isenthalpic depressurization from P1 to P2*

This calculation is peculiar to liquid–vapor mixtures.

The aim is to calculate V and T so that $H_2 = H_1$.

The mixture in question must compulsorily contain liquid unless certain components are in the supercritical state, in which case retrograde condensation may occur.

This détente – i.e. a drop in pressure – can take place through a valve, following a relatively long path through a pipe.

The enthalpy of the mixture is of the form:

$$H(V,(x),(y),T,P)$$

The fed mixture is defined by:

$$H_1(V_1,(x)_1,(y)_1,T_1,P_1)$$

The aim is to calculate V_2, $(x)_2$, $(y)_2$ and T_2 so that:

$$H_2 = H_1 \qquad \text{and} \qquad P = P_2$$

In reality, V is a function of T through the "N" equation. The same is true for x_i and y_i, so we can simply write:

$$H = H(T, P_2)$$

It is therefore sufficient to zero the function:

$$F(T) = H(T) - H_1$$

The initial state will simply be the state of the fed mixture but taken at the pressure P_2 and temperature T_1.

NOTE.– Asselineau *et al.* [ASS 79] proposed to use the *global method* of the Jacobian (see Appendix 6) to solve the above nonlinear equation systems.

5.3.9. *The "N" equation in the presence of non-condensable gaseous inerts*

The standardization of the molar fractions in the overall mixture is written:

$$\sum_{1}^{n} z_i + z_I = 1$$

where z_I is the molar fraction of inerts.

As before:

$$z_i = (1 - V)x_i + VE_i x_i$$

$$x_i = \frac{z_i}{1 - V + E_i V} \quad \text{and} \quad y_i = \frac{E_i x_i}{1 - V + E_i V}$$

In the present case:

$$\sum_{1}^{n} x_i = 1 \qquad \sum_{1}^{n} y_i + y_I = 1 \qquad y_I = \frac{z_I}{V}$$

The "N" function is now written as:

$$N(T, P, V) = \sum_{1}^{n} x_i - 1 = \sum_{1}^{n} \frac{z_i}{1 - V + E_i V} - 1 = 0$$

5.3.10. *Two condensable vapors and an inert*

In this case, it is possible to directly determine the vapor as a function of the temperature without iterations by resolution of a simple second-degree equation.

Let us write that:

$$\sum_{1}^{2} x_i = 1$$

$$1 = \frac{z_1}{1 - V + E_1 V} + \frac{z_2}{1 - V + E_2 V}$$

This is a second-degree equation of the form:

$$aV^2 + bV + c = 0$$

with:

$$a = +(1 - E_1)(1 - E_2)$$

$$b = -(2 + E_1 z_2 + E_2 z_1 - z_1 - z_2 - E_1 - E_2)$$

$$c = 1 - z_1 - z_2 = z_1$$

From this, we deduce the root that needs to be used:

$$V = \frac{-b - \sqrt{b^2 - 4ac}}{2a}$$

NOTE.– In the absence of inerts: $c = 0$ and $V = -b/a$

with: $z_1 + z_2 = 1$

$$a = (1 - E_1)(1 - E_2)$$

$$b = (1 + E_1 z_2 + E_2 z_1 - E_1 - E_2)$$

5.3.11. *Mixture of water and hydrocarbons*

The liquid is now the union of two phases and, because of their low mutual solubility, we obtain:

– water containing a few per mil of hydrocarbons;

– an organic phase containing a few per mil of water.

From the point of view of the liquid–vapor equilibrium, then, as an initial approximation, we can consider that the water contains no hydrocarbons and that the organic phase contains no water.

For the water, the equilibrium equation is:

$$x_e = 1 \qquad\qquad y_e = E_e x_e \# E_e$$

For the hydrocarbons, it is:

$$y_i = E_i x_i$$

The global material balance is written thus:

$$V + L_e + L_h = 1 \text{ (index h for "hydrocarbons")}$$

The partial balances become:

$$z_i = x_i(1 - V - L_e) + VE_i x_i$$

$$z_e = L_e + VE_e \qquad \text{hence} \qquad L_e = z_e - VE_e$$

and:

$$z_i = x_i(1 - V - z_e + VE_e + VE_i)$$

The N function is then written:

$$N(T, P, V) = \sum_1^n x_i - 1 = \sum_1^n \frac{z_i}{1 - V - z_e + V(E_e + E_i)} - 1 = 0$$

5.4. A limiting case: Henry's law

5.4.1. *Henry's law*

Henry's law is used for low concentrations in the liquid phase.

For $x_i \rightarrow 0$:

$$H_i \longrightarrow \gamma_{i\infty} \pi_i(T) = \left[\lim \frac{f_i}{x_i}\right]_{x_i \rightarrow 0}$$

In order for Henry's law to be valid, in the vicinity of $x = 0$, the value H_i *must depend only on the temperature* and not on the composition – i.e. on x_i:

$$H_i(T) = \gamma_{i\infty} \pi_i(T) \quad \text{with} \quad \gamma_{i\infty} = \text{const.} \quad \text{for} \quad x_i \# 0$$

The sign ∞ represents "infinite dilution".

Often the temperature T is greater than the critical temperature T_{ci}, such that $\pi_i(T)$ is simply a pseudo vapor pressure. Naturally, nothing stands in the way of γ_i varying with the temperature.

For high pressures and at ambient temperature, the fugacity of the gas is identical for both phases, and can be assimilated to the pressure of the gas above the solvent – often water, which has very low volatility in these conditions, and we write:

$$P = Hx_s$$

For gases such as O_2, N_2 and H_2 in solution in water, P is of the order of 100 bars, and the molar fraction x_s of the dissolved gas, for its part, is of the order of 10^{-3}.

5.4.2. *Influence of the pressure on the solubility of a gas*

As variables, let us take the pressure and the composition:

$$d\mathrm{Lnf}_i^x = \left(\frac{\partial \mathrm{Lnf}_i^x}{\partial P}\right)_z dP + \left(\frac{\partial \mathrm{Lnf}_i^x}{\partial \mathrm{Lnz}_i}\right)_P d\mathrm{Lnz}_i$$

The exponent x may be either L (liquid) or V (vapor).

For a gas $(i = 2)$ that is not highly soluble in a solvent $(i = 1)$:

$$\left(\frac{\partial Lnf_2^L}{\partial Lnx_2}\right)_{P,T} = 1 \quad (\gamma_2 = \text{const.})$$

Thus:

$$dLnf_2^L = \left(\frac{\partial Lnf_2^L}{\partial P}\right)_{x_2} dP + dLnx_2 = \frac{\overline{v^L}}{RT} dP + dLnx_2$$

The volatility of the solvent is supposed to be low, so that, in the gaseous phase:

$$y_2 \# 1 \quad Lny_2 = 0 \quad \text{and} \quad dLny_2 = 0$$

Hence:

$$\frac{dLnx_2}{dP} = \frac{\partial Lnf_2^V}{\partial P} - \frac{\partial Lnf_2^L}{\partial P} = \frac{\overline{v_2^V} - \overline{v_2^L}}{RT} = \frac{\Delta v}{RT}$$

In addition, at equilibrium (which persists if P varies):

$$f_2^L = f_2^V \quad \text{and} \quad dLnf_2^L = dLnf_2^V$$

When the pressure varies, $\overline{v_2^L}$ hardly varies at all, whilst $\overline{v_2^G}$ varies greatly.

For:

$-$ Low P $\Delta v > 0$ and x_2 grows with P;

$-$ High P $\Delta v < 0$ and x_2 decreases when P increases.

In other words, x_2 passes through a maximum.

5.5. Liquid–liquid equilibrium: general points and representations

5.5.1. *Terminology*

In liquid–liquid extraction, two situations may arise:

1) When we refine a *noble product N*, the solvent dissolves the *impurity I* and the raffinate R rich in N and poor in I remains.

2) When we extract the *noble species* N, the solvent leaves the *residue* R (rich in I, poor in N and very poor in solvent S).

We agree that:

– the index P characterizes the species I or N which is hardly, or is not, dissolved by the solvent;

– the index T characterizes the species transferred I or N and that, consequently, the solvent greatly dissolves;

– the index S characterizes the solvent.

Regarding the liquid phases, by convention:

– the framework E characterizes the liquid rich in solvent – i.e. the extract. In this liquid, the molar fractions will be y_P, y_T and y_S;

– the framework R characterizes the residue or the raffinate. The molar fractions will be x_P, x_T and x_S;

– the framework F characterizes the feed in question. The molar fractions will be z_P, z_T and z_S.

The feed F may be:

– either a homogeneous mixture of N and I. Such is the case, generally, with the feeding of extractors;

– or a mixture of two immiscible liquids, each containing all three species P, T and S. It is this latter situation which is of interest in the study of liquid–liquid equilibria. This study consists of seeking the proportions of the two liquids E and R (when $E + R = 1$) and their composition.

5.5.2. *Barycentric coordinates*

Consider a mixture of three components P, T and S. We define that mixture by its total "weight" (expressed in kmoles) and the molar fractions of the three components. Similarly, in geometry, we define a vector by its modulus and its directive cosines. Below, mixtures are represented with a star in the same way as vectors are represented with an arrow.

We use the notion of "weight" which, here, will be the number of kilomoles present in a liquid or a mixture.

By combining various mixtures, we form a mixture M^* with the "weight" M and whose composition is given by the three fractions corresponding to the three components.

Writing that the mixture M is obtained by the union of the two compounds A and B is equivalent to writing the equations:

$$M = A + B \qquad \text{Overall balance}$$

$$\left. \begin{array}{l} Mx_T^M = Ax_T^A + Bx_T^B \\ Mx_P^M = Ax_P^A + Bx_P^B \\ Mx_S^M = Ax_S^A + Bx_S^B \end{array} \right\} \text{ Partial balances}$$

We shall agree to replace the above set of equations with the following formula:

$$M^* = A^* + B^*$$

It is possible to attach a geometrical meaning to the above. Any given triangle (but usually an isosceles or equilateral triangle) can be used to represent the composition of a mixture C^* with three components T, P and S.

Let A be the point of intersection of the straight lines TC^* and PS. The ratio AC^*/AT defines the titer x_T of the component T in the mixture C^*.

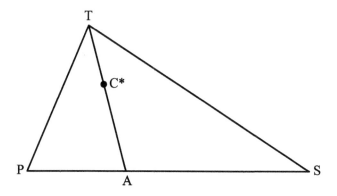

Figure 5.7. *Triangular diagram (principle)*

We see that it is possible, in advance, to plot three sets of lines parallel to the three sides of the triangle, with those lines being numbered, for example, from 0 to 1 in intervals of 0.1. Such diagrams are found in commerce.

In these conditions, elementary geometrical considerations show that:

1) $x_T + x_P + x_S = 1$

2) Consider three mixtures A, B and C whose "weights" are arbitrary and whose images A^*, B^* and C^* are aligned. The ratio of the lengths of the straight-line segments AB and AC is written:

$$\frac{AB}{AC} = \frac{x_i^A - x_i^B}{x_i^A - x_i^C} \text{ (for i between 1 and 3)}$$

3) Now consider the mixtures A and B with weights M_A and M_B, whose union is the mixture R with weight $M_R = M_A + M_B$, and let x_i^A, x_i^B and x_i^M represent the fractions of the compound i (i between 1 and 3) in each of the mixtures. We can write:

$$(M_A + M_B)x_i^R = M_A x_i^A + M_B x_i^B$$

meaning that:

$$M_A(x_i^R - x_i^A) = -M_B(x_i^R - x_i^B)$$

and finally, regarding the straight-line segments RA and RB:

$$\frac{M_A}{M_B} = -\left[\frac{x_i^R - x_i^B}{x_i^R - x_i^A}\right] = -\frac{RB}{RA}$$

The image R^* is therefore the barycenter of A^* and B^*, assigned the weights M_A and M_B, with those weights being numbers of kilomoles.

4) The point C^*, which is the image of the compound C^*, is the barycenter of the three points T, P and S, assigned the weights x_T, x_P and x_S. This can be verified graphically.

5.5.3. Liquid–liquid equilibrium in the triangular representation

Liquid–liquid equilibrium is usually studied at constant temperature (and at constant pressure).

In order for solvent-based extraction to be possible, it is necessary for the diagram to have a domain of immiscibility. In the representation in Figure 5.8, this domain is delimited by the concave curve which is based on the straight line PS. This curve includes a critical point C. The arc to the left of C (on the side of P) represents the compositions of the raffinate (or residue) R and the arc to the right of C (on the side of S) represents the extracts E. The conjugation lines RE link the raffinate and the extract at mutual equilibrium. Any mixture situated outside of the curve (called the demixing curve) is homogeneous.

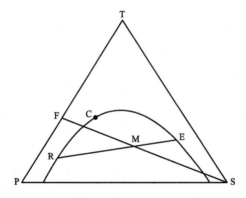

Figure 5.8. *Ternary equilibrium*

When the raffinate and the extract move away from the line PS, they eventually meet at the critical point C, where the conjugation line has a length of zero.

5.5.4. A simplification: the sharing curve

If the mutual solubility of S and P remains low in the working domain, we can discount it and only take into consideration the solubility of T in S and P – i.e. in E and R.

We let y represent the molar fraction of T in E, and x the molar fraction of T in R. Extraction is favored if $y > x$. The sharing curve is often concave toward the x axis. It is imprudent to continue the sharing curve right up to the critical point because, then, we can no longer disregard the mutual solubility of S and P.

The sharing curve is helpful in graphically determining the theoretical number of stages of a counter-flow extraction using a construction similar to McCabe and Thiele's for distillation (see [DUR 16]).

5.5.5. An extreme simplification: the sharing coefficient

If the quantity of solute to be extracted from the raffinate (or the residue) is slight, we can often assimilate the sharing curve to its tangent to the origin and define the sharing coefficient m by:

$$y/x = const. = m$$

This hypothesis leads to fairly simple analytical solutions. The calculations use the efficacy parameter ε which is defined by:

$$\varepsilon = m \frac{S}{P} = m \frac{E}{R} = \frac{yE}{xR}$$

Extraction will be easier when m and S/P (or E/R) have high values – in other words, when the parameter ε is large.

The efficacy ε is the ratio between the kmoles of solute present in the extract and the raffinate. The sharing-coefficient method, even approximate, is very useful to initialize a calculation of liquid–liquid equilibrium.

5.5.6. *Particular case of petroleum-based oils*

One possible procedure is as follows, to characterize the extraction of the aromatic compounds present in an oil.

The oil to be de-aromatized is brought into contact with an excess of solvent (the furfural) in the pure state and repeat the operation 2 to 4 times. We strip the solvent and re-mix the extracts obtained. This mixture will be the essentially aromatic part A of the oil. However, the "raffinate" – i.e. the oil having undergone the extractions – after stripping of the little dissolved solvent, represents the essentially paraffinic fraction P of the oil being analyzed.

It is then possible to establish a traditional ternary diagram in solvent-based extraction by mixing the solvent S and the fractions A and P in variable proportions.

Having measured the equivalence in theoretical stages of an extractor with simple products (solvent, benzene, heptane), it is then possible, by graphical construction, to predict the results of the processing of an oil.

A similar procedure is envisageable for the extraction of the aromatics from a catalytic reformate or for the de-asphalting of a propane bitumen.

5.6. Calculation of liquid–liquid equilibria: binary mixtures.

5.6.1. *Description of the binary equilibrium (2 phases and 2 components)*

The molar Gibbs energy is written as follows, whatever the number of components:

$$g = g_o + g^M + g^E$$

For two components (see Figure 5.9):

$$g_o = x_1 g_{o1}(T, P) + (1 - x_1) g_{o2}(T, P).$$

At the index, o represents a mean value obtained by weighting with the molar fractions.

$$g^M = RT\left[x_1 Lnx_1 + (1 - x_1) Ln(1 - x_1)\right]$$

g^E is the excess molar Gibbs energy. For example, according to Renon [REN 68, REN 71]:

$$g^E = RT\sum_{i=1}^{2} x_i \frac{\displaystyle\sum_{j=1}^{2} x_j G_{ji} \tau_{ji}}{\displaystyle\sum_{k=1}^{2} x_k G_{ki}}$$

where:

$$G_{ji} = \exp(-\alpha_{ji} \tau_{ji}) ; \ \tau_{ji} \neq \tau_{ij} ; \ \tau_{ii} = 0 ; \ 0.2 < \alpha_{ji} = \alpha_{ij} < 0.5$$

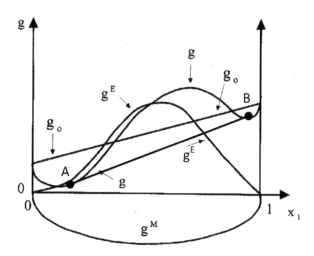

Figure 5.9. *Shape of variations with* x_1 *of* $g = g_o + g^E + g^M$

g^E is not symmetrical with respect to the two components of a binary mixture. In addition, $g^E = 0$ if x_1 or x_2 is null. The function g must accommodate two inflection points, which is possible only if g^E also has two. Such is the case of Renon's function g^E for certain values of the parameters.

5.6.2. *The common tangent rule (binary mixtures)*

The Gibbs energy of a binary mixture is:

$$G = (n_1 + n_2)g(x_1) = n_T g(x_1) \text{ where } n_T = \text{const.}$$

The chemical potential of compound 1 is:

$$\mu = \frac{\partial G}{\partial n_1} = \frac{\partial G}{\partial x_1} \times \frac{\partial x_1}{\partial n_1}$$

However:

$$x_1 = \frac{n_1}{n_T} \text{ and } \frac{\partial x_1}{\partial n_1} = \frac{\partial G}{\partial x_1} = n_T \frac{\partial g(r)}{\partial x_1}$$

and:

$$\frac{\partial G}{\partial n_1} = \frac{n_T}{n_T} \times \frac{\partial g(x)}{\partial x_1} = \frac{\partial g(x)}{\partial x}$$

Finally, the chemical potential of compound 1 is:

$$\mu_1 = \frac{\partial G}{\partial n_1} = \frac{\partial g(x_1)}{\partial x_1} = g'(x_1) = \mu$$

However, the equilibrium between the A and B phase is written as follows for compound 1:

$$\mu_A = \mu_B = \mu$$

$$g = x_1\mu \quad \text{and} \quad g_A = x_{1A}\mu \quad \text{and} \quad g_B = x_{1B}\mu$$

$$g_A - g_B = (x_A - x_B)\mu$$

Finally:

$$\mu = \frac{dg}{dx}\bigg|_A = \frac{dg}{dx}\bigg|_B = \frac{g_A - g_B}{x_A - x_B}$$

In Figure 5.9, *the points A and B are the points of contact of the curve g(x) with a common tangent AB.*

5.6.3. Calculation of the equilibrium of a binary mixture

If there are only two components, we know that:

$$\left(\frac{dg}{dx_1}\right)_A = \frac{g(x_A) - g(x_B)}{x_A - x_B} = \left(\frac{dg}{dx_1}\right)_B$$

To calculate the derivatives, we may, at will:

– establish the mathematical form of $g' = dg/dx$ and use it throughout the calculations;

– give x_1 a variation $\Delta x_1 = 10^{-5}$ and deduce the resulting variation Δg.

We can see that the direction of the variation of g with x_1 changes sign if the difference $(g_{n-1} - g_{n-2})$ is of opposite sign to that of $(g_n - g_{n-2})$. Therefore, we can be sure that the function g has passed through an extremum in the interval $[x_{n-2}, x_n]$.

The equilibrium is calculated as follows:

1) We divide the interval $(0,1)$ of variation x_1 into 50 equal intervals. Out of the three intervals A, I and B where the sign of g' changes, we discount the interval corresponding to x_1 such that:

$$x_A < x_I < x_B$$

We take the two extreme intervals and name $(x_{j-2}, x_j)_A^{(0)}$ the "interval j" and $(x_{k-2}, x_k)_B^{(0)}$ the "interval k".

2) We calculate:

$$\overline{x_k} = \frac{1}{2}(x_{k-2} + x_k) \quad \text{and} \quad \overline{x_j} = \frac{1}{2}(x_{j-2} + x_j)$$

Thus:

$$\frac{g(\overline{x_j}) - g(\overline{x_k})}{\overline{x_j} - \overline{x_k}} = p^{(0)}$$

3) We divide the two intervals k and j by 20 and, for each sub-interval, we calculate:

$$X^{(0)} = g' - p^{(0)}$$

In the interval k, we note the value x_m corresponding to a change of sign of X. We do the same for the sub-intervals j, which gives x_ℓ. Thus, we obtain the sub-intervals $(x_{m-1}, x_m)^{(1)}$ and $(x_{\ell-1}, x_\ell)^{(1)}$.

4) We calculate:

$$\overline{x_m} = \frac{1}{2}(x_{m-1} + x_m) \quad \text{and} \quad \overline{x_\ell} = \frac{1}{2}(x_{\ell-1} + x_\ell)$$

and: $p^{(1)} = \dfrac{g(\overline{x_m}) - g(\overline{x_\ell})}{\overline{x_m} - \overline{x_\ell}}$

5) If $\dfrac{p^{(n)} - p^{(n-1)}}{p^{(n)}} < 10^{-3}$, then the calculation is complete.

Otherwise, we return to 3 with new sub-intervals.

5.7. Ternary mixtures

A ternary mixture contains three components and one or two phases.

Here, we shall discuss only those mixtures which contain a single zone of demixing, as indicated in Figure 5.10.

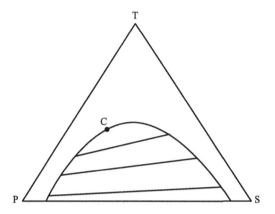

Figure 5.10. *Ternary diagram and correspondence lines*

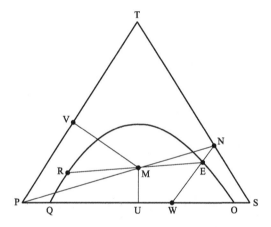

Figure 5.11. *Approximate content of E in terms of solvent*

The principle of the calculation is to quickly determine the approximate compositions of the liquids R and E created by the decomposition of the initial mixture M. Unlike what happens with liquid–vapor mixtures, there is no "ideal law" to estimate the equilibrium ratios E_i. Therefore, we need a graphical approach.

5.7.1. *Multiple binary demixings*

Two or three binary demixings define ternary demixing zones as represented in Figure 5.12.

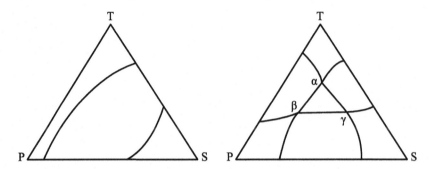

Figure 5.12. *Ternary diagrams containing 2 and 3 binary demixings*

5.8. Equilibria in the supercritical domain

5.8.1. *Supercritical two-phase equilibria*

In order for such an equilibrium to be possible, it is necessary for the mixture to contain at least one fluid in the supercritical state. The gaseous mixture is, at pressure, in the dense state, which means that its density is of the same order of magnitude as that of a liquid. In this state, the solvent gas can selectively dissolve a component that we are seeking to isolate. Therefore, it is sufficient to slightly depressurize the "gaseous solution" for the solute to separate (preferably in the divided solid state). The gas is then recompressed for a cycle.

There are also couples of gases that are immiscible when the pressure of system is greater than the critical pressures of the two gases in question.

At the pressure of the system, there is a temperature above which there is no longer any demixing. This temperature increases with pressure.

5.8.2. *Transfer of material in the critical domain*

In the vicinity of the critical conditions, the two phases present tend toward a common state, which leads the gradients of the chemical potential to approach zero. In addition, the diffusivities of material tend to be cancelled out.

In these conditions, the transfer of material becomes unpredictable.

5.8.3. *Supercritical extraction procedures*

The principle of the process is to dissolve a substance with low volatility in a fluid in the supercritical state. In these conditions, we need to aim for maximum solubility. Solubility increases with the density of the solvent and with temperature. Indeed, a high value of the critical temperature of the gas is a positive factor.

In the same way as a liquid selectively absorbs a component from a gaseous mixture, a gas may selectively strip (entrain) a component from a liquid mixture. To improve the effectiveness of the stripping, we can add a so-called "stripper" fluid to the supercritical gas. A good stripper should increase the dependency of the solvent power on the temperature and pressure. A stripper with a critical temperature lower than that of the solvent can reduce the solubility of certain components that are not highly volatile, and thus improve the selectivity of the extraction.

To carry out the extraction, there are two types of procedures available to us: compression or pumping. If the extraction pressure is moderate (100 bars, for example), the pumping cycle consumes more energy (for heat exchange) than the compression cycle. The reverse is true when the pressure is high (e.g. 1000 bars), because the electrical energy needed to drive the compressor is significant.

Carbon dioxide is commonly used in high-pressure extractions. The critical characteristics of that gas are:

$$T_c = 304.15 \text{ K} \qquad\qquad P_c = 73.8 \text{ bar}$$

NOTE (Calculation of supercritical equilibria).– We shall not give a general method here. Quite simply, we need to find a compromise between the liquid–vapor methods and liquid–liquid methods. Brunner's work offers numerous commented examples of equilibrium between fluids. Finally, this type of calculation may very well employ an equation of state.

APPENDICES

Appendix 1

Indefinite Integrals Useful for the Calculation of the Functions of State on the Basis of a Cubic Equation

A1.1. Expression of the "cubic" equation

The function of state is supposed to be of the following form:

$$P = \frac{RT}{V-b} - \frac{a}{V^2 + \beta V + \gamma}$$

A1.2. Expressions of the integrals

$$I_1 = \int \frac{dV}{V^2 + \beta V + \gamma} = \frac{1}{\sqrt{\beta^2 - 4\gamma}} \, Ln \left[\frac{2V + \beta - \sqrt{\beta^2 - 4\gamma}}{2V + \beta + \sqrt{\beta^2 - 4\gamma}} \right]$$

$$I_2 = \int \frac{dV}{(V^2 + \beta V + \gamma)^2} = \frac{2V + \beta}{(4\gamma - \beta^2)(V^2 + \beta V + \gamma)} + \frac{2I_1}{4\gamma - \beta^2}$$

$$I_3 = \int \frac{VdV}{(V^2 + \beta V + \gamma)^2}$$

$$I_3 = -\frac{\beta V + 2\gamma}{(4\gamma - \beta^2)(V^2 + \beta V + \gamma)} - \frac{\beta I_1}{4\gamma - \beta^2}$$

$$I_4 = \int \frac{dV}{(V^2 + \beta V + \gamma)^3}$$

$$I_4 = \frac{3}{16\left(\dfrac{\beta^2}{4} - \gamma\right)^{5/2}} \, \mathrm{Ln}\left[\frac{2V + \beta - \sqrt{\beta^2 - 4\gamma}}{2V + \beta + \sqrt{\beta^2 - 4\gamma}}\right]$$

$$+ \frac{3(V + \beta/2)}{8\left(\dfrac{\beta^2}{4} - \gamma\right)^2 (V^2 + \beta V + \gamma)} - \frac{V + \beta/2}{(\beta^2 - 4\gamma)(V^2 + \beta V + \gamma)^2}$$

Solving Third- and Fourth-degree Equations: Searching for Dimensioless Groups

A2.1. Third-degree equation

Consider the cubic equation:

$$x^3 + \alpha x^2 + \beta x + \gamma = 0$$

We set:

$$p = \beta - \frac{\alpha^2}{3}$$

$$q = \gamma - \frac{\alpha\beta}{3} + \frac{2\alpha^3}{27}$$

Thus:

1) If $\dfrac{q^2}{4} + \dfrac{p^3}{27} > 0$

$$x = -\frac{\alpha}{3} + \left[\frac{-q}{2} + \sqrt{\frac{q^2}{4} + \frac{p^3}{27}} \right]^{1/3} + \left[\frac{-q}{2} - \sqrt{\frac{q^2}{4} + \frac{p^3}{27}} \right]^{1/3}$$

2) If $\dfrac{q^2}{4} + \dfrac{p^3}{27} < 0$

$$y_1 = 2\sqrt{\dfrac{-p}{3}}\cos\dfrac{a}{3}: \quad y_2 = 2\sqrt{\dfrac{-p}{3}}\cos\left(\dfrac{a}{3} + \dfrac{2\pi}{3}\right): \quad y_3 = 2\sqrt{\dfrac{-p}{3}}\cos\left(\dfrac{a}{3} + \dfrac{4\pi}{3}\right)$$

where:

$$a = \text{Arc}\cos\dfrac{3q}{\rho p}, \quad \rho^2 = \dfrac{-4p}{3}: \quad x_i = y_i - \dfrac{\alpha}{3}$$

A2.2. Fourth-degree equation

The most typical form of a fourth-degree equation is:

$$x^4 + d_3 x^3 + d_2 x^2 + d_1 x + d_0 = 0$$

Let us set:

$$\delta = d_3/2 \qquad \alpha = 2A - \delta^2$$

$$A = d_2 - \delta^2 \qquad \beta = A^2 + 2B\delta - 4d_0$$

$$B = d_1 - A\delta \qquad \gamma = -B^2$$

Now consider the cubic equation:

$$\xi^3 + \alpha\xi^2 + \beta\xi + \gamma = 0$$

Because the left-hand side of this equation is negative when $\xi = 0$, the equation certainly has a positive root ξ:

We then set:

$$\delta = d_3/2 \qquad \varepsilon = \dfrac{1}{2}(A + \xi)$$

$$d = \sqrt{\xi} \qquad e = \dfrac{d}{2}\left(\delta - \dfrac{B}{\xi}\right)$$

and the solutions to the fourth-degree equation are the same as those to the two second-degree equations:

$$x^2 + (\delta + d)x + (\varepsilon + e) = 0$$

$$x^2 + (\delta - d)x + (\varepsilon - e) = 0$$

A2.3. Finding dimensionless groups expressing a physical law

Consider a physical law involving n physical properties, expressed by k fundamental values. For example, the physical property of velocity is the quotient of the length by time.

Buckingham showed that the number of dimensionless groups is:

$$i = n - k$$

On page 349 of [BUC 14], he gives an example of the calculation of the i dimensionless groups.

However, the author shows that the number of systems of dimensionless groups possible (each containing i groups) is equal to i. We then need to turn to the physics of the phenomena in order to choose the appropriate system.

From page 356 onwards of [BUC 14], the author surrenders to considerations on the units of electromagnetism. Today, though, those considerations are deemed outdated.

With regard to the finding of the dimensionless groups, readers could usefully consult [BUC 14] and [BUC 15].

Appendix 3

A Few Important Identities

A3.1. Theorem of reciprocity

We suppose that:

$$F(x,y) = \text{const.} = C \quad \text{and} \quad dF = 0$$

The values of x and y are not independent, and we can always write:

$$y = f(x) \quad \text{and} \quad dy = \left(\frac{dy}{dx}\right)_F dx$$

$$dF = \left(\frac{\partial F}{\partial x}\right)_y dx + \left(\frac{\partial F}{\partial y}\right)_x dy = \left[\left(\frac{\partial F}{\partial x}\right)_y + \left(\frac{\partial F}{\partial y}\right)_x \left(\frac{\partial y}{\partial x}\right)_F\right] dx = 0$$

The value in square brackets must be equal to zero:

$$\left(\frac{\partial F}{\partial x}\right)_y = -\left(\frac{\partial F}{\partial y}\right)_x \left(\frac{\partial y}{\partial x}\right)_F$$

[A3.1]

Similarly, if we were to switch x and y around, we would have:

$$\left(\frac{\partial F}{\partial y}\right)_x = -\left(\frac{\partial F}{\partial x}\right)_y \left(\frac{\partial x}{\partial y}\right)_F$$

[A3.2]

By multiplying, taking each side of the equation in turn, we obtain the theorem of reciprocity:

$$\left(\frac{\partial y}{\partial x}\right)_F \left(\frac{\partial x}{\partial y}\right)_F = 1 \text{ or indeed } \left(\frac{\partial y}{\partial x}\right)_F = \frac{1}{\left(\dfrac{\partial x}{\partial y}\right)_F} \qquad [A3.3]$$

A3.2. Closed-loop derivation

In view of equations [A3.1] and [A3.3]:

$$\left(\frac{\partial F}{\partial y}\right)_x \left(\frac{\partial y}{\partial x}\right)_F \left(\frac{\partial x}{\partial F}\right)_y = -1$$

In a closed loop, the numerator of the first derivative is equal to the denominator of the third derivative.

A3.3. Cascaded derivation

We suppose that:

$$F = f(x) \text{ and } x = g(u)$$

The cascaded derivation is:

$$\frac{dF}{du} = \left(\frac{dF}{dx}\right)\left(\frac{dx}{du}\right)$$

which can be written thus (in light of the theorem of reciprocity):

$$\left(\frac{dF}{dx}\right)_u \left(\frac{dx}{du}\right)_F \left(\frac{du}{dF}\right)_x = 1$$

NOTE.– There is no contradiction between closed-loop derivation and cascaded derivation. Indeed:

– in closed-loop derivation, one of the variables is constant and its total differential is zero;

– in cascaded derivation, and for two derivatives out of the three, each variable is an explicit function of the next. The third derivative can be deduced from this.

We can see the importance of the dependencies and constraints being clearly defined.

A3.4. Rational fractions and simple elements

Before performing any derivation, and particularly before any integration, it is *imperative* to break down the rational fractions into simple elements. For example:

$$\frac{1}{g_2} = \frac{1}{(V-V_1)(V-V_2)} = \frac{1}{(V_1-V_2)}\left[\frac{1}{(V-V_1)} - \frac{1}{(V-V_2)}\right]$$

$$\frac{1}{g_3} = \frac{1}{(V-V_1)(V-V_2)(V-V_3)} = \frac{1}{D_{12} \times D_{13}(V-V_1)}$$
$$+ \frac{1}{D_{23} \times D_{21}(V-V_2)} + \frac{1}{D_{31} \times D_{32}(V-V_3)}$$

with:

$$D_{ij} = (V_i - V_j)$$

Appendix 4

A Few Expressions for Partial Vapor Pressures

A4.1. Henry's constant

Henry's law is expressed by:

$$p = Hx$$

p: vapor pressure of the solute at equilibrium,

x: molar fraction of the solute,

p is measured in atmospheres.

H is therefore measured in atmospheres, and is dependent on the temperature. Thus, for ammonia dissolved in water at a content of less than 15%:

$$H = \exp\left(12.186 - \frac{3031.16}{t + 229.35}\right)$$

H in atm

t in°C

When using Henry's law, it must be remembered that, in parallel, the vapor pressure of the solvent varies in accordance with Raoult's law – i.e. it is proportional to the molar fraction of the solvent.

A4.2. Empirical formulae

In certain cases – e.g. when we have complexation in solution or even a chemical reaction – it is necessary to make use of empirical expressions.

Such is the case with the dissolution of ammonia gas and hydrochloric gas in water (hereinafter, x represents mass fractions).

1) NH_3–water system: Concentrations less than 15%: use Henry's law.

NH_3: concentrations greater than or equal to 15%, use the following relation:

$$p = \exp\left[\left(11.01672 - \frac{2814.219}{t+230}\right)\right]\bigg/\left(\frac{1-1.6x}{x}\right)\left(0.7574 + \frac{62.51}{t+230}\right)$$

Water:

$$p = \exp\left[13.778154 + 0.267373x - 2.74972x^2\right.$$

$$\left. -\left(\frac{5185.22 - 52.3998x + 156.448x^2}{t+273}\right)\right]$$

2) HCl–water system:

HCl:

$$p = \exp\left[20.24525 - 14.31977x - \left(\frac{11121.269 - 22170.193x + 17972.592x^2}{t+273}\right)\right]$$

Water:

$$p = \exp\left[13.88267 - 1.0834x - \left(\frac{5326.137 - 21262.392x + 7290.398x^2}{t + 273}\right)\right]$$

For both systems:

− p is expressed in atmospheres;

− x is expressed in *gravimetric* fractions.

Numerical Calculation
of the Solution to F(X) = 0

Suppose we know an interval encapsulating the sought solution and where F(X) is monotonic. If this is not the case, then based on considerations of order of magnitude, we choose an interval which we divide into 50 elementary intervals and, for each of these, we ensure that:

– for i between 0 and 50 $(F_{i+1} - F_i) (F_i - F_{i-1}) > 0 \Rightarrow$ monotonic nature around the section I;

– for $i = k$ $F_k \times F_{k-1} < 0 \Rightarrow$ existence of a solution in the interval k.

If there are multiple possible solutions, we disregard those which make no physical sense.

Having thus chosen the interval (between $k-1$ and k), we numerically calculate the derivative – i.e. the slope of the string:

$$F'\#m^{(0)} = \frac{F(X_k + \varepsilon X_k) - F(X_k)}{\varepsilon X_k} \quad (\text{where } \varepsilon = 10^{-7})$$

$$X^{(1)} = X^{(0)} - \frac{F(X^{(0)})}{m^{(0)}}$$

and, more generally:

$$m^{(n)} = \frac{F(X^{(n)}) - F(X^{(n)} - \varepsilon X^{(n)})}{\varepsilon X^{(n)}} \quad \text{and} \quad X^{(n+1)} = X^{(n)} - \frac{F(X^{(n)})}{m^{(n)}} \quad \text{where}$$

$\varepsilon = 10^{-7}$.

The calculation is halted when:

$$X^{(n+1)} - X^{(n)} < 10^{-5} X^{(n)}$$

Appendix 6

Jacobian Method

The $2n+3$ variables u_i which define a thermodynamic system are linked by $2n+3$ equations [ASS 79]. We have seen that two of these equations are tantamount to imposing the value of two variables of state.

– the vector (U), whose u_i are the coordinates;

– the vector (G), whose coordinates are the values G_k of the left-hand sides of the equations;

– the Jacobian of the system. The Jacobian is a matrix [J] whose most general element is:

$$j_{ki} = \frac{\partial G_k}{\partial u_i}$$

Symbolically, we write:

$$[J] = \left[\frac{\partial G}{\partial u} \right]$$

Based on an initial estimation of the variables u_i, for these variables we need to try to find increases Δu_i which bring the G_k closer to zero. We write:

$$(G)^{(n)} + [J]^{(n)} (\Delta U)^{n+1} = 0$$

and, if $[K]^{(n)}$ is the inverse matrix of $[J]^{(n)}$, we have:

$$(\Delta U)^{(n+1)} = -[K]^{(n)}(G)^{(n)}$$

Convergence is obtained when the modulus of (ΔU) is less than 10^{-12}.

From one iteration to the next:

– the variables must remain positive;

– the density of the liquid must be greater than that of the vapor;

– the modulus of $(G)_{n+1}$ must not be greater than that of $(G)_n$.

If this is not the case, (ΔU) is divided by 2, up to five times if necessary.

The Jacobian method can be described as universal in that, at the cost of inverting the matrix, it avoids interlocking iterative loops, but the looped method more clearly reflects the physical reality, and may help prevent false solutions.

Appendix 7

Characteristics of Various Gases

Gas	Molar mass	γ (15°C)	P_{cri}	T_{cri}
Acetylene	26	1.24	62.4	172
Air	29	1.40	37.7	133
Ethyl Alcohol	46	1.13	63.9	517
Methyl Alcohol	32	1.20	79.8	513
Ammoniac	17	1.31	114.2	406
Argon	40	1.66	48.6	151
Nitrogen	28	1.40	33.9	127
Benzene	78	1.12	49.2	563
Isobutane	58	1.10	36.5	408
n-Butane	58	1.09	38	425
Isobutylene	56	1.10	40	418
Butylene	56	1.11	40.2	420
Chlorine	71	1.36	77.1	417
Ethyl Chloride	64.5	1.19	52.7	460
Methyl Chloride	50.5	1.20	66.7	417
n-Decane	142	1.03	22.1	619

Carbon Dioxide	44	1.30	74	304
Sulfur Dioxide	64	1.24	78.7	430
Ethane	30	1.19	48.8	305
Ethylene	28	1.24	51.2	283
Water gas	19.5	1.35	31.3	130
Catalytic Cracker gas	28.8	1.20	46.5	286
Coke Oven gas	10.7	1.35	29.1	109
Blast Furnace gas	29.6	1.39	–	–
Natural gas	18.8	1.27	46.5	210
Helium	4	1.66	2.3	5
n-Heptane	100	1.05	27.4	540
n-Hexane	86	1.06	30.3	508
Hydrogen	2	1.41	13	33
Methane	16	1.31	46.4	191
Carbon Monoxide	28	1.40	35.2	134
n-Nonane	128	1.04	23.8	596
n-Octane	114	1.05	24.9	569
Oxygen	32	1.40	50.3	154
Isopentane	72	1.08	33.3	461
n-pentane	72	1.07	33.7	470
Propane	44	1.13	42.5	370
Propylene	42	1.15	46	365
Hydrogen Sulfide	34	1.32	90	374
Steam	18	1.33	221.2	648

Appendix 8

The CGS Electromagnetic System

A8.1. Potential in the international system (SI) and in the CGS electromagnetic system

By definition (if $1C = 1$ coulomb)

$$1\underset{SI}{C} = 3.10^9 q_{CGSEM} \text{ and } \varepsilon_0 = \frac{1}{36\pi.10^9} C.m^{-1}.V^{-1}$$

The potential created in a vacuum by a 1 C charge at a distance of 1 m is:

$$Pot_{SI} = \frac{1C}{4\pi\varepsilon_0 1m} = 9.10^9 Volt$$

The potential created by a charge q of 1 CGSEM unit at the distance of 1 cm is:

$$Pot_{CGSEM} = \frac{q}{1cm} = 1 \text{ volt stat}$$

The ratio between these two potentials must be such that:

$$\frac{Pot_{SI}}{Pot_{EMCGS}} = \frac{C}{q} \times \frac{cm}{m}$$

Put differently:

$$\frac{9.10^9 Volt}{1 \text{ volt stat}} = \frac{C}{q} \frac{cm}{m} = \frac{3.10^9}{100}$$

Thus:

$$1\,V = \frac{1\,\text{volt stat}}{300}$$

Note that we can very simply write:

$$\text{Volt} \times C = \text{Joule}$$

$$\text{Volt stat} \times q = \text{erg}$$

Therefore:

$$\frac{1\,\text{Volt}}{1\,\text{volt stat}} = \frac{\text{Joule}}{\text{erg}} \times \frac{q}{C} = 10^7 \times \frac{1}{3.10^9} = \frac{1}{300}$$

A8.2. Other units in SI and CGSEM

1) The conductivity κ is measured in $\Omega^{-1}.m^{-1}$. The conductance is the quotient of the intensity by the potential.

Finally:

$$[\kappa] = \frac{I}{VL} = \frac{C}{V.s.m} \qquad (\text{SI})$$

However:

$$1\,C = 3.10^9\,q_{CGSEM}$$

$$1\,V = \frac{1}{300}\,\text{volt stat}$$

$$1\,m = 100\,\text{cm}$$

Hence:

$$\kappa_{SI} = \frac{3.10^9 \times 300}{100}\,\kappa_{CGSEM} = 9.10^9\,\kappa_{CGSEM}$$

2) Although viscosity is not an electromagnetic property, its correspondence between the CGS system and the SI is given here. Viscosity is measured in Pa.s, which is to say $kg.m^{-1}.s^{-1}$.

However:

1 kg = 1000 g

1 m = 100 cm

Therefore:

1 Pa.s = 10 g. $cm^{-1}.s^{-1}$ = 10 poise = 10 barye.s

Resistance, Conductance, Diffusance

1) For electricity, Ohm's law is written:

$$\Delta V = Ri = i \times \left(\frac{\rho \Delta L}{S}\right) = \text{electrical flow through surface S} \times \text{resistance}$$

The intensity is measured in Coulomb.s^{-1}.

The resistance expression features:

– the resistivity, ρ (Ω.m);

– the distance traveled, ΔL;

– the cross-section, S.

2) In heating technology, Fourier's law is written:

$$q = \lambda \times \frac{\Delta T}{\Delta L} S \quad (\text{J.s}^{-1} = \text{Watt})$$

or:

$$\Delta T = q \times \left(\frac{\Delta L}{S\lambda}\right) = \text{heat flow through surface S} \times \text{conductance}$$

The conductance expression features:

– the conductivity, λ (J.m^{-1}.°C^{-1}.s^{-1});

– the distance traveled, ΔL;

– the cross-section, S.

3) For the diffusion of materials, Fick's law is written:

$$m = SD\frac{\Delta C}{\Delta L} \ (\text{kg. s}^{-1} \text{ or kmol. s}^{-1})$$

or:

$$\Delta C = m \times \left(\frac{\Delta L}{SD}\right) = \text{flow of matter through surface S} \times \text{diffusance}$$

C: concentration in kg or kmol per m^3

The diffusance expression features:

– diffusivity D ($m^2.s^{-1}$),

– the distance traveled, ΔL,

– the cross-section, S.

In these three laws, we see that the temperature T is a thermal (i.e. heat) potential, and the concentration C is a material potential. V is evidently the electrical potential, i.e. the voltage (in Volts).

NOTE.– The following grouping may be used in thermal calculations:

$$\alpha = \frac{\lambda}{\rho C_p} \ \text{ with dimensions } \ \frac{J}{s\,m\,°C} \times \frac{m^3}{kg} \times \frac{kg\,°C}{J} = \frac{m^2}{s}$$

Grouping α is referred to as thermal diffusivity, as it uses the same units as the material diffusivity D.

Bibliography

[ABR 84] ABRAHAM M.H., "Thermodynamics of solution of homologous series of solutes in water", *J. Chem. Soc. Faraday Trans.*, vol. 80, p. 153, 1984.

[AND 80a] ANDERSON T.F., PRAUSNITZ J.M., "Computational methods for high-pressure phase equilibria and other fluid-phase properties using a partition function 1-Pure fluids", *Ind. and Eng. Chem. Process Des. Dev.* vol. 19, pp. 1–8, 1980.

[AND 80b] ANDERSON T.F., PRAUSNITZ J.M., "Computational methods for high-pressure phase equilibria and other fluid-phase properties using a partition function 2-Mixtures", *Ind. and Eng. Chem. Process Des. Dev.* vol. 19, pp. 9–14 1980.

[ARC 90] ARCHER D.G., PEIMING W., "The dielectric constant of water and Debye-Hückel limiting law slopes", *J. Phys. Chem. Ref. Data*, vol. 19, no. 2, p. 371, 1990.

[ASE 99] ASEYEV G.G., *Electrolytes equilibria in solutions and phase equilibria* Editions Begell House Inc., 1999.

[ASS 79] ASSELINEAU L., BOGDANIC G., VIDAL J., "A versatile algorithm for calculating vapor-liquid equilibria", *Fluid phase equilibria*, vol. 3, pp. 273–290, 1979.

[BAL 82] BALIAN R., *Du microscopique au macroscopique*, Editions Ellipses 1982.

[BAZ 89] BAZAROV I., *Thermodynamique*, Editions Mir, 1989.

[BEE 84] BEERBOWER A., WU P.L., MARTIN A., "Expanded solubility parameter approach. I: naphthalene and benzoic acid in individual solvents", *J. of Pharma Sciences*, vol. 73, no. 2, p. 179, 1984.

[BEN 70] BENDER E., "Equations of state exactly representing the phase behavior of pure substances", *Proc. 5th Syump. Thermophys. Progr. ASME*, New York, p. 227, 1970.

[BJE 29] BJERRUM N., "Neuere Anchauungen über Electrolyte", *Berichte d. D. Chem. Gesellschaft*, vol. 62, p. 1091, 1929.

[BRO 73] BROMLEY L.A., "Thermodynamic properties of strong electrolytes in aqueous solutions", *A.I. Ch. E. Journal*, vol. 19, p. 313, 1973.

[BRU 94] BRUNNER G., *Gas Extraction*, Springer, 1994.

[BUC 14] BUCKINEHAM E., "On physically similar systems. Illustrations of the use of dimensional equations", *Physical Review*, vol. 4, no. 4, pp. 345–376, 1914.

[BUC 15] BUCKINEHAM E., "The principle of similitude", *Nature*, vol. 96, pp. 396–397, 1915.

[CAB 81] CABANI S., GIANNI P., MOLLICA V. *et al.*, "Group contribution to the thermodynamic properties of non-ionic organic solutes in dilute aqueous solution", *J. of Solution Chemistry*, vol. 10, p. 563, 1981.

[CAL 58] CALDERBANK P.H., "Physical rate processes in industrial fermentation – Part. I The interfacial area in gas-liquid contacting with mechanical agitation", *Trans. Inst. Chem. Engrs.*, vol. 36, p. 443, 1958.

[CAL 59] CALDERBANK P.H., "Physical rate processes in industrial fermentation. Part II Mass transfer coefficients in gas-liquid contacting with and without mechanical agitation", *Trans. Inst. Chem. Engrs.*, vol. 37, p. 173, 1959.

[CAL 61] CALDERBANK P.H., "Physical rate processes in industrial fermentation Part III Mass transfer from fluids to solid particles suspended in mixing vessels", *Trans. Inst. Chem. Engrs.*, vol. 39, pp. 363–368, 1961.

[CHI 34] CHILTON T.H., COLBURN A.P., "Mass transfer (absorption) coefficients Ind.", *Eng. Chem.*, vol. 26, p. 1183, 1934.

[CHU 68] CHUEH P.L., PRAUSNITZ J.M., "Calculation of high-pressure vapor-liquid equilibria", *Ind. and Eng. Chemistry*, vol. 60, no. 3, pp. 34–52, 1968.

[CRU 78] CRUZ J.-L., RENON H., "A new thermodynamic representation of binary electrolyte solutions nonideality in the whole range of concentrations", *A.I. Ch. E. Journal*, vol. 24, p. 817, 1978.

[CRU 79] CRUZ J.-L., RENON H., "Nonideality in weak binary electrolytic solutions. Vapor-liquid equilibrium data and discussion of the system water-acetic acid", *Ind. Eng. Chem. Fundam.*, vol. 18, p. 168, 1979.

[CUS 84] CUSSLER E.L., *Diffusion. Mass transfer in fluid systems*, Cambridge University Press, 1984.

[DAN 80a] DANESI P.R., CHIARIZIA R., "The kinetics of metal solvent extraction", *CRC. Critical Reviews in Analytical Chemistry*, vol. 10, no. 1, pp. 1–125, 1980.

[DAN 80b] DANESI P.R., CHIARIZIA R., VAN DE GRIFT G.F. "Kinetics and mechanism of the complex formation reactions between Cu (II) and Fe (III) aqueous species and a β-Hydroxy Oxime in toluene", *Journal of Physical Chemistry*, vol. 84, no. 25, pp. 3455–3461, 1980.

[DAN 80c] DANESI P.R., VAN DE GRIFT G.F., HORWITZ E.P. *et al.*, "Simulation of interfacial two-step consecutive reactions by diffusion in the mass-transfer kinetics of liquid-liquid extraction of metal cations", *Journal of Physical Chemistry*, vol. 84, no. 26, pp. 3582–3587, 1980.

[DEB 23] DEBYE P., HÜCKEL E., "Zur Theorie der Electrolyte", *Phys. Zeitschr.*, vol. 9, p. 185, May 1923.

[DUR 53] DURAND E., *Electrostatique et magnetostatique*, Masson, 1953.

[DUR 16] DUROUDIER J.-P., *Distillation*, ISTE Press, London and Elsevier, Oxford, 2016.

[EDM 61] EDMISTER W.C., *Applied Hydrocarbon Thermodynamics*, Gulf Publishing Company, 1961.

[EIN 06] EINSTEIN A., "Eine neue Bestimmung der Moleküldimensionen", *Annalen der Physik (Leipzig)*, vol. 19, p. 289, 1906.

[EYR 64] EYRING H., HENDERSON D., JONES STOVER B., *Statistical Mechanics and Dynamics,* John Wiley and Sons, 1964

[FIC 55] FICK A., "Ueber Diffusion", *Annalen der Physik*, vol. 170, p. 59, Leipzig, 1855.

[FLO 41a] FLORY P.J., "Constitution of three-dimensionel polymers and the theory of gelation", *Eighteenth Colloid Symposium*, Cornell University, New York, June 19–21 1941.

[FLO 41b] FLORY P.J., "Thermodynamics of high polymer solutions", *J. Chem. Phys.*, vol. 9, p. 660, 1941.

[FLO 41c] FLORY P.J., "Thermodynamics of high polymer solutions", *J. Chem. Phys.*, vol. 10, p. 51, 1941.

[FLO 44] FLORY P.J., "Thermodynamics of heterogeneous polymers and their solutions", *J. Chem. Phys.*, vol. 12, no. 11, p. 425, 1944.

[FRE 77] FREDENSLUND A., GMEHLING J., RASMUSSEN P., *Vapor-Liquid Equilibria Using UNIFAC, a Group-Contribution Method*, Elsevier, 1977.

[FUL 66] FULLER E.N., SCHETTLER P.D., GIDDINGS J.C., "A new method for prediction of binary gas-phase diffusion coefficients", *Ind. Eng. Chem.*, vol. 58, p. 19, 1966.

[GIB 99] GIBBS J.W., *Equilibre des systèmes chimiques,* Gauthier-Villars, 1899

[GIL 34] GILLILAND E.R., "Diffusion coefficients in gaseous systems", *Ind. Eng. Chem.*, vol. 26, p. 681, 1934.

[GLA 71] GLANSDORFF P., PRIGOGINE I., *Structure, stabilité et fluctuations*, Masson, 1971.

[GME 79] GMEHLING J., LIU D.D., PRAUSNITZ J.M., "High-pressure vapor-liquid equilibria for mixtures containing one or more polar components" *Chem. Eng. Science*, vol. 34, p. 951, 1979.

[GME 82] GMEHLING J., RASMUSSEN P., FREDENSLUND A., "Vapor-liquid equilibria by UNIFAC group contribution. Revision and extension", *Ind. Eng. Chem. Process Des. Dev.*, vol. 21, p. 118, 1982.

[GRE 07] GREEN D.W., PERRY R.H., *Perry's Chemical Engineers' Handbook*, 8th Edition, McGraw Hill, 2007.

[GRE 99] GREINER W., NEISE L., STÖCKER H., *Thermodynamique et mécanique statistique*, Springer, 1999.

[GUI 49] GUINIER G., *Eléments de physique moderne théorique,* Editions Bordas, 1949.

[HEI 80] HEIDEMANN R.A., KHALIL A.M., "The calculation of critical points", *A.I. Ch. E. Journal*, vol. 26, no. 5, p. 769, 1980.

[HEL 74] HELGESON H.C., KIRKHAM D.H., "Theoretical prediction of the thermodynamic behavior of aqueous electrolytes at high pressures and temperatures: I. Summary of the thermodynamic electrostatic properties of the solvent", *Am. Journal of Science*, vol. 274, p. 1089, 1974.

[HEL 81] HELGESON H.C., KIRKHAM D.H., FLOWERS G.C., "Theoretical prediction of the thermodynamic behavior of aqueous electrolytes at high pressures and temperatures IV. Calculation of activity coefficients, osmotic coefficients, and apparent molal and standard and relative partial molal properties to 600°C and 5 kbar", *Am. Journal of Science*, vol. 281, p. 1289. 1981.

[HIG 35] HIGBIE R., "The rate of absorption of a pure gaz into a still liquid during short periods of exposure", *Trans. Am. Inst. of Chem. Engrs*, vol. 31, p. 365, 1935.

[HIL 29] HILDEBRAND J.H., "Solubility XII Regular solutions" *J.Am.Chem. Soc.*, vol. 51, p. 66, 1929.

[HIL 47] HILDEBRAND J.H., "The entropy of solution of molecules of different sizes", *J. Chem. Physics*, vol. 15, no. 5, p. 225, 1947.

[HUG 41] HUGGINS M.L., "Solutions of long chain compounds", *J. of Chemical Physics*, vol. 9, p. 440, 1941.

[JEN 81] JENSEN T., FREDENSLUND A., RASMUSSEN P., "Pure-component vapor-pressures using UNIFAC group contribution" *Ind. Eng. Chem. Fundam*, vol. 20, p. 239, 1981.

[KER 50] KERN D.Q., *Process Heat Transfer*, McGraw-Hill, 1950.

[KIT 61] KITTEL C., *Eléments de physique statistique*, Dunod, 1961.

[KRI 77] KRISHNA R., "A generalised film model for mass transfer in non-ideal fluid mixtures", *Chem. Eng. Science*, vol. 32, p. 659, 1977.

[KRI 78] KRISHNA R., "A thermodynamic approach to the choice of alternatives to distillation", *I. Chem. E. Sympos. Series no. 54*, 1978.

[KRI 79] KRISHNA R., STANDART G.L., "Mass and energy transfer in multicomposent systems", *Chem. Eng. Communic*, vol. 3, p. 201, 1979.

[KUS 73] KUSIK C.L., MEISSNER H.P., "Vaper pressure of water over aqueous solutions of strong electrolytes", *Ind. Eng. Chem. Process Des. Dev.*, vol. 12, p. 112, 1973.

[KUS 78] KUSIK C.L., MEISSNER H.P., "Electrolyte activity coefficients in inorganic processing", *A.I.Ch. E. Symp. Series*, vol. 74, p. 14, 1978.

[LEE 75] LEE B.I., KESLER M.G., "A generalized thermodynamic correlation based on three-parameter corresponding states" *A.I. Ch. E. Journal* vol. 21, no. 3, p. 510 1975.

[LEN 93] LENCKA M., RIMAN R.E., "Thermodynamic modeling of hydrothermal syntheses of ceramic powders", *Chem. Mater.,* vol. 5, p. 61, 1993.

[LI 97] LI J., CARR P.W., "Accuracy of empirical correlations for estimating diffusion coefficients in aqueous organic mixtures", *Analytical Chemistry*, vol. 69, pp. 2530–2536, 1997.

[LON 69] LONCIN M., *Die grundlagen der verfahrenstechnik in der lebensmittelindustrie*, Sauerländer, Aarau and Frankfurt, 1969.

[LON 85] LONCIN M., *Génie industriel alimentaire*, Masson, 1985.

[LYK 61] LYKLEMA J., OVERBECK J., TH.G., "On the interpretation of electrokinetic potentials", *Journal of Colloïd Science*, vol. 16, pp. 501–512, 1961.

[MAN 71] MANSOORI G.A., CARNAHAN N.F., STARLING K.E. *et al.*, "Equilibrium thermodynamic properties of the mixture of hard spheres", *Journal of Chemical Physics*, vol. 54, p. 1523, 1971.

[MAN 90] MANZO R.H., AHUMADA A.A., "Effects of solvent medium on solubility. Part V: Enthalpic and entropic contribution to the free energy changes of disubstitued benzene derivatives in ethanol: water and ethanol cyclohexane mixtures", *J. of Pharma. Sciences*, vol. 79, no. 12, p. 1109, 1990.

[MEI 72a] MEISSNER H.P., KUSIK C.L., "Activity coefficients of strong electrolytes in multicomponent aqueous solutions", *A.I.Ch.E. Journal*, vol. 18, p. 294, 1972.

[MEI 72b] MEISSNER H.P., TESTER J.W., "Activity coefficients of strong electrolytes in aqueous solutions", *Ind. Eng. Chem. Process. Des. Dev.*, vol. 11, p. 128, 1972.

[MEI 73a] MEISSNER H.P., KÜSIK C.L., "Aqueous solutions of two or more strong electrolytes", *Ind. Eng. Chem. Process Des. Dev.*, vol. 12, p. 205, 1973.

[MEI 73b] MEISSNER H.P., PEPPAS N.A., "Activity coefficients. Aqueous solutions of polybasic acids and their salts", *A. I. Ch. E. Journal*, vol. 19, p. 806, 1973.

[MEI 79] MEISSNER H.P., KÜSIK C.L., "Double salt solubilities", *Ind. Eng. Chem. Process Des. Dev.,* vol. 18, p. 391, 1979.

[MEI 80] MEISSNER H.P., "Prediction of activity coefficients of strong electrolytes in aqueous systems", *Am. Chem. Soc. Symp. Series 133*, vol. 495, 1980.

[MIC 80] MICHELSEN M.L., "Calculation of phase envelopes and critical points for multicomponent mixtures", *Fluid Phase Equilibria*, vol. 4, p. 1, 1980.

[MIC 81] MICHELSEN M.L., HEIDEMANN R.A., "Calculation of critical points from cubic two-constant equations of state", *A.I. Ch. E. Journal*, vol. 27, no. 3, p. 521, 1981.

[MIL 69] MILAZZO G., *Electrochimie,* Dunod, 1969.

[MOO 72] MOORE W.J., *Physical Chemistry,* Prentice Hall, 1972.

[NAB 90] NABETAMI H., NABAJIMA M., WATARNABE A. *et al.*, "Effects of osmotic pressure and adsorption on ultrafiltration of ovalbumine", *A.I.Ch.E.J.*, vol. 96, p. 907, 1990.

[NAG 78] NAGATA I., OHTA T., "Prediction of the excess enthalpies of mixing of mixtures using the UNIFAC method", *Chemical Engineering Science*, vol. 33, p. 177, 1978.

[NEL 54] NELSON L.C., OBERT E.F., "Generalised PVT properties of gases", *Trans. of the American Soc. of Mechanical Engineers*, vol. 76, p. 1057, 1954.

[NOU 85] NOUGIER J.P., *Méthodes de calcul numérique*, Masson, 2nd edition, 1985.

[OCO 67] O'CONNELL J.P., PRAUSNITZ J.M., "Empirical correlation of second virial coefficients for vapor-liquid equilibrium calculations I.E.C.", *Process Design and Development* vol. 6, p. 245, 1967.

[ONS 31a] ONSAGER L., "Reciprocal relations in irreversible processes I", *Physical Review,* vol. 37, p. 405, 1931.

[ONS 31b] ONSAGER L., "Reciprocal relations in irreversible processes II", *Physical Review,* vol. 38, p. 2265, 1931.

[ORB 98] ORBEY, *Modeling Vapour Liquid Equilibria: Cubic Equations of State and their Mixing Rules*, Cambridge University Press, 1998.

[PAR 59] PARIS A., *Procédés de rectification*, Dunod, 1959.

[PEN 76] PENG D.Y., ROBINSON D.B., "A new two-constant equation of state", *Ind. Eng. Chem. Fundam,* vol. 15, no. 1, p. 59, 1976

[PEN 77] PENG D.Y., ROBINSON D.B., "A rigorous method for predicting the critical properties of multicomponent systems from an equation of state", *A.I. Ch. E. Journal*, vol. 23, no. 2, p. 137, 1977

[PER 36] PERRIN F., "Mouvement brownien d'un ellipsoïde", *Le Journal de physique et Le Radium*, vol. 7, p. 1, 1936.

[PIT 73a] PITZER K.S., MAYORGA G., "Thermodynamics of electrolytes. II Activity and osmotic coefficients for strong electrolytes with one or both ions univalents", *The Journal of Physical Chemistry*, vol. 77, p. 2300, 1973.

[PIT 73b] PITZER K.S., "Thermodynamics of electrolytes. I Theoretical basis and general equations", *Journal of Physical Chemistry*, vol. 77, p. 268, 1973.

[PIT 74] PITZER K.S., MAYORGA G., "Thermodynamics of electrolytes. III Activity and osmotic coefficients for 2-2 electrolytes", *Journal of Solution Chemistry,* vol. 3, p. 539, 1974.

[PRA 86] PRAUSNITZ J.M., LICHTENTHALER R.N., GOMES DE AZEVEDO E., *Molecular Thermodynamics of Fluid-Phase Equilibria*, 2nd edition, Prentice Hall, 1986.

[PRI 68] PRIGOGINE I., *Introduction à la thermodynamique des processus irreversible*, Dunod, 1968.

[PRI 99] PRIGOGINE I., KENDEPUDI D., *Thermodynamique*, Editions Odile Jacob, 1999.

[REI 66] REID R.C., SHERWOOD T.K., *The Properties of Gases and Liquids*, McGraw Hill, 1966.

[REI 85] REIF F., *Statistical and Thermal Physics*, McGraw Hill, 1985.

[REN 68] RENON H., PRAUSNITZ J.M., "Local compositions in thermodynamic excess functions for liquid mixtures", *A.I.Ch.E Journal*, vol. 14, no. 1, p. 135, 1968.

[REN 71] RENON H., ASSELINEAU L., COHEN G. *et al.*, *Calcul sur ordinateur des équilibres liquide-vapeur et liquide-liquide*, Editions Technip, 1971.

[SCA 70] SCATCHARD G., RUSH R.M., JOHNSON J.S., "Osmotic and activity coefficients for binary mixtures of sodium chloride sodium sulfate magnesium sulphate and magnesium chloride in water at 25°C. III Treatment with the ions as components", *Journal of Physical Chemistry*, vol. 74, p. 3780, 1970.

[SCH 46] SCHRÖDINGER E., *Statistical Thermodynamics*, Cambridge University Press, 1946.

[SCH 64] SCHEIBEL E.G., "Liquid diffusivities", *I.E. Chem.*, vol. 46, p. 2007, 1964.

[SCH 90] SCHWARTZENTRUBER J., RENON H., "Values for nonideal system. An easier way", *Chemical Engineering*, p. 118, March 1990.

[SHO 88] SHOCK E.L., HELGESON H.C., "Calculation of the thermodynamic and transport properties of aqueous species at high pressures and temperatures: correlation algorithm for ionic species and equation of state prediction to 5 kbar and 1000°C", *Geochimica et Cosmochimica Acta*, vol. 52, p. 2009, 1988.

[SHO 89] SHOCK E.L., HELGESON H.C., SVERJENSKY D.A., "Calculation of the thermodynamic and transport properties of aqueous species at high pressures and temperatures: standard partial molal properties of inorganic neutral species", *Geochimica et Cosmochimica Acta*, vol. 53, p. 2157, 1989.

[SHO 90] SHOCK E.L., HELGESON H.C., "Calculation of the thermodynamic and transport properties of aqueous species at high pressures and temperatures: standard partial molal properties of organic species", *Geochimica et Cosmochimica Acta*, vol. 54, p. 915, 1990.

[SLA 72] SLATTERY J.C., *Momentum, energy and mass transfer in continua*, McGraw Hill, 1972.

[STI 98] STICHLMAIR J.G., FAIR J.R., *Distillation, principles and practice*, Wiley-VCH, 1998.

[STO 67] STOUGHTON R.W., LIETZE M.H., "Thermodynamic properties of sea salt solutions", *J. Chem. Eng. Data*, vol. 12, p. 101, 1967.

[TAN 88] TANGER J.C., HELGESON H.C., "Calculation of the thermodynamic and transport properties of aqueous species at high pressures and temperatures: revised equations of state for the standard partial molal properties of ions and electrolytes", *American Journal of Science*, vol. 288, p. 19, 1988.

[TAY 80] TAYLOR R., WEBB D.R., "Stability of the film model for multicomposent mass transfert", *Chem. Eng. Commun*, vol. 6, p. 175. 1980.

[TAY 81] TAYLOR R., WEBB D.R., "Film model for multicomposent mass transfert: computational methods: the exact solution of the Maxwell–Stefan equations", *Computers and Chemical. Eng.*, vol. 5, p. 61, 1981.

[TRE 80] TREYBAL R.E., *Mass Transfer Operations*, McGraw Hill, 1980.

[VID 73] VIDAL J., *Thermodynamique*, Editions Technip, 1973.

[VID 74] VIDAL J., *Thermodynamique, Méthodes appliquées au raffinage et au génie chimique*, Editions Technip, 1974.

[VIG 66] VIGNES A., "Diffusion in binary solutions", *I.E.C. Fund.*, vol. 5, p. 189, 1966.

[VON 29] VON BORN M., "Volumen und Hydratationswärme der Iouen", *Zeitschrift Physik 1*, vol. 45, 1929.

[WIL 55] WILKE C.R., CHANG P., "Correlation of diffusion coefficients in dilute solutions", *A.I.Ch.E. J.*, vol. 1, p. 264, 1955.

[WIL 64] WILSON G.M., "Vaper-liquid equilibrium XI. A new expression for the excess free energy of mixing", *J. Am. Chem. Soc.*, vol. 86, p. 127, 1964.

[WON 92] WONG D.S.H., SANDLER S.I., "A theoretically correct mixing rule for cubic equations of state", *A.I. Ch. E. Journal*, vol. 38, no. 5, p. 671, 1992.

[WU 68] WU Y.C., RUSH R.M., SCATCHARD G., "Osmotic and activity coefficients for binary mixtures of sodium chloride, sodium sulphate, magnesium sulphate and magnesium chloride in water at 25°C. I Isopiesty measurements on the four systems with common ions", *Journal of Physical Chemistry,* vol. 72, p. 4048, 1968.

[WU 69] WU Y.C., RUSH R.M., SCATCHARD G., "Osmotic and activity coefficients for binary mixtures of sodium chloride, sodium sulphate, magnesium sulphate and magnesium chloride in water at 25°C. II Isopiesty and electromotive force measurements on the two systems without common ions", *Journal of Physical Chemistry*, vol. 75, p. 2047, 1969.

[WUI 65] WUITHIER P., *Raffinage et génie chimique*, Editions de l'I.F.P, 1965.

[YOU 80] YOUNG M.E. CARROAD P.A., BELL R.L., "Estimation of diffusion coefficients of proteins", *Biotechn. and Bioengineering*, vol. 22, p. 947, 1980.

Index